# Raising Sheep

*An Essential Guide on How to Raise Sheep in Your Backyard or on a Small Farm*

# Contents

# Introduction

Raising sheep can be a rewarding venture. People consider raising sheep in their backyard or small farms for many reasons; you may raise sheep as a hobby, to gain financial benefits, or to become more self-sufficient. This book aims to provide you with all the information that small-scale sheep farming requires to raise sheep in your backyard or small farm successfully, despite your aspirations and expectations.

Like most other livestock, raising sheep comes with many challenges, especially on a small-scale. This book provides simple, practical, and valuable tips, instructions, advice, and guidelines for those planning to raise sheep and those who already own small herds. The book addresses such common obstacles that you're highly likely to face when raising sheep, including overcoming diseases, predators, securing financial viability, and addressing legal and social issues.

The amount of space you have should not be a major concern as long as your herd's size suits the available space. The book aims to help small-scale farmers best-utilize the limited space available to ensure an exciting and rewarding sheep-raising venture.

Most individuals and families who live in urban and suburban areas opt not to raise sheep for a variety of reasons. Some incorrectly believe that raising sheep requires perfect pastureland, but sheep consume many types of grasses, weeds, and brush that can grow almost anywhere. You do not need perfect pastureland to raise a small flock of sheep. This book provides all the right tips to raise a happy herd of sheep in almost any land, including land in urban or suburban areas.

Sheep are relatively easier to manage compared to other livestock such as pigs, cattle, horses, llamas, and donkeys. They are usually gentle and calm except for rams. They can be easily trained to follow verbal cues, just like most other farm animals. Raising sheep is much easier than you might think.

You may be worried about the size of land you own, but raising a small herd of sheep with five ewes and their lambs only requires approximately an acre of land. In return, you and your family will get to enjoy raising sheep, interacting with them, their organic meat, milk, wool, and excellent manure that can be used in home gardens.

Different breeds of sheep are more or less suited for different purposes. It's important that you pick the right breed to suit your needs and expectations. The needs of sheep, such as food, water, pasture, and shelter, vary between breeds. It's important to pick the right breed when raising them in a backyard or small farm. The book provides valuable information on different breeds of sheep so you can choose the most suitable breed for your venture.

Raising sheep in your backyard or small farm requires an initial investment, but it is usually much less than the investment required to raise other popular types of livestock. The book aims to show you how much of an investment you will need to raise sheep.

This book can be a tool to help determine whether you should raise sheep by understanding the basic needs of sheep, the benefits of raising sheep, the commitment it takes, challenges that need to be overcome, and the investment and costs associated with it.

Furthermore, it will come in handy as a guide you can refer to throughout the venture, especially when you're faced with obstacles.

# Chapter 1: Facts About Sheep and Raising Them

Sheep-raising dates back thousands were years since sheep were one of the first animals to be domesticated by humans. People raise sheep for many reasons. Foremost, they provide wool, meat, and milk that satisfy basic human needs. They are easy to raise, and they can withstand harsh climates when provided with protection from predators. They remain one of the most popular types of livestock globally, with more than one billion domesticated sheep throughout the globe.

According to Debra K. Aaron and Donald G. Ely at the University Of Kentucky College Of Agriculture, the first evidence of domesticated sheep comes from Central Asia, 11,000 years ago. There are over 900 breeds of sheep around the world, and there are 50 breeds in the United States. According to Aaron and Ely, there are approximately five million sheep in the United States, with Texas, California, Colorado, Wyoming, and Utah being the top five states for raising sheep.

Sheep belong to the same family that livestock animals such as goats, cattle, and antelopes belong to, known as Bovidae. They are ruminants like goats and cattle, which means their stomachs feature four separate compartments. They re-chew partly digested plant matter they eat, known as cud. It's known as rumination that stimulates digestion by further breaking down plant matter.

Adult male sheep fully intact and not castrated are known as rams and bucks. Castrated sheep are known as wethers. Adult female sheep are called ewes. Sheep that are less than a year are lambs. A group of sheep is usually called a flock. They are considered grazers because they prefer vegetation such as grasses, brush, and legumes that grow closer to the ground. Sheep are gregarious, which means they like to stay close to each other in flocks. They are easier to move and look after.

Sheep raised for wool are usually shorn once every year. The wool shorn from a single sheep is called a Fleece, while the wool shorn from a flock is a clip. According to WorldAtlas.com, Australia is the world's biggest wool producer. Australia produced 25% of the global wool production in 2015-2016, helping it become approximately a three-billion-dollar industry.

Sheep raised for meat have fewer wool fibers and more hair fibers. They shed their coats every year, and do not need shearing. Lambs are usually sold for meat when they weigh around 90 to 130 pounds. China is the world's leading producer of lamb by some distance, with Australia and New Zealand ranked 2nd and 3rd.

# Benefits of Raising Sheep in Backyards and Mini-Farms

Small-scale sheep farming offers many benefits. Different individuals and families may be attracted to raising sheep in their backyards and small properties for different reasons. People may be motivated by the profits that sheep farming offers by utilizing land that isn't used for any other purpose. Wool, meat, and milk can make households more self-sufficient and help them make significant savings, while excess produce can generate profits. Small-scale farming also enables households to enjoy tax concessions in a few states.

Besides being financially attractive, sheep farming is also highly enjoyable. While sheep are easy to handle due to their gentle nature, they make great pets when socialized with humans when they are lambs. Raising sheep involves many daily chores perfect for families that intend to live more outdoorsy lifestyles.

Many people who raise sheep on a smaller scale do it as a beloved pastime. Some small-scale farmers find motivation in maintaining expensive breeding stock, while others may find satisfaction in achieving high production goals. For example, a small-scale sheep farmer may be the owner of a great ram with a great breeding value. Another farmer may find achieving a 200% lamb crop as a great success, while another may find satisfaction in avoiding losses during the lambing season.

Love and concern for the environment that people live in is also another motivator for raising sheep on a small scale. The practice offers organic meat and milk much eco-friendlier and has a smaller carbon footprint. The wool that sheep provide can not only be sold, but also, it's used to make clothes. Sheep manure can be used as organic fertilizer in home gardens.

### Economic Benefits

Raising sheep usually is mainly to generate an income for the individual, family, or farm they belong to. While sheep are the primary income of farms, others consider it a secondary or a minor revenue stream since sheep farming complements other agricultural ventures well. Raising sheep in backyards and mini-farms offers financial benefits, although it usually can't be considered a primary income source.

Many sheep need to be farmed for the income to be sufficient for an individual or family, but a backyard or mini-farm can only accommodate a few sheep. As a result, revenue is usually smaller.

Most small-scale farmers consider the revenue from sheep farming as a secondary income source.

## Wool

Although the apparel industry has greatly evolved over the past few decades, wool is still being largely used to make clothing, bedding, furniture, and even insulation for homes. As a matter of fact, many industry experts project that wool will retake its throne from synthetic materials as the world becomes more environmentally conscious.

Wool is considered one of the best fibers for insulation, according to Good Shepherd Wool. Moreover, it is non-carcinogenic, flame resistant, sustainable, recyclable, blocks sounds, and absorbs toxins. Most other fabrics used for insulation do not offer such a versatile mix of beneficial attributes.

Those who raise sheep in their backyards or small farms can enjoy a good profit by selling wool. A pound of wool can be sold for at least a dollar, while a fleece can bring revenue of at least $10. Sheep are sheared once every year, usually in the spring. The wool should then be inspected for dirt and other debris before being sold.

## Meat

The market for lamb and mutton is steadily increasing. People are discovering the benefits that eating sheep meat provides in contrast to beef and chicken. As a result, there is a good demand for lamb and mutton. Those who farm sheep in their backyards or mini-farms can take their sheep to a slaughterhouse or the market itself.

There is a high demand for lamb and mutton, especially lamb, at local farmer's markets. People who like to consume organic meats with a smaller carbon footprint and restaurants that use such ingredients often use farmer's markets to source fresh and organic lamb and mutton. As a result, small-scale sheep farmers can enjoy higher profits by selling lamb and mutton at local farmer's markets.

## Milk

Needless to say that sheep's milk isn't as popular as cow's milk or goat's milk. But experts believe that sheep milk offers many nutritional benefits. According to nutritionists, sheep milk has twice as much Calcium as cow milk does. It also has high levels of vitamin C, B, B12, riboflavin, and thiamin that help boost the body's nervous and immune systems.

Sheep milk is more of a staple in a few areas of the world. It is commonly used to make various cheeses such as Roquefort, feta, and ricotta. Sheep milk can yield more cheese per ounce of fluid as it contains more solids in it compared to cow milk. Sheep milk can also produce yogurt.

Anyone who wishes to raise sheep on a small scale for milk needs to go for a dairy breed. Dairy breeds such as Lacaune and East Friesian usually produce double the amount of milk than non-dairy breeds. A healthy dairy breed ewe can produce approximately 1,000 pounds of milk every year. Sheep farmers can sell sheep milk, cheese, and yogurt at farmer's markets or in bulk to restaurants to enjoy substantial revenue.

## Breeding

Sheep that come from good bloodlines are in high demand. Breeding sheep is a profitable venture, especially for small-scale farmers since they rarely have enough land to accommodate large flocks. Health, fertility, mothering ability, feed efficiency, and production are the most important factors that breeders need to focus on.

There is a growing market for sheep with good genetics in foreign markets. Exporting sheep can be lucrative for small-scale farmers, but sheep need to be in very good health for export, and farmers need to satisfy various guidelines set by importing countries. Sheep with good bloodlines can also be sold in local markets to turn sizable profits.

## Tax Concessions

People with enough land are motivated to raise sheep since it makes them entitled to tax concessions. Different states offer different concessions for farmers, even the small-scale ones. Raising sheep makes people entitled to lower property taxes at agricultural rates if the property meets the definition of a "farm" under the particular state's guidelines.

Even agricultural enterprises are subject to various taxes, such as income tax. Although, raising livestock such as sheep makes people entitled to tax write-offs. In such cases, common farm expenditures such as capital purchases can be written off against income since most sheep-related purchases are exempted from sales taxes.

## Manure

Sheep manure is highly fertile. It can be used in home gardens instead of commercial fertilizer to enjoy healthy and organic produce. A small flock of sheep can easily produce enough manure required for a sizable home garden that includes vegetables, herbs, fruits, and flowers.

## Easy Maintenance of Land

Maintaining a property that is a few acres large takes quite a lot of time and effort. Sheep can be looked at as living lawnmowers that don't cost time and money. They love all kinds of vegetation, including weeds. Also, letting sheep graze the land to keep it tidy is eco-friendlier than using lawnmowers.

Farmers rent out flocks of goats to clear out unwanted vegetation such as brush from properties. Sheep farmers can rent out small flocks of sheep to individuals and businesses to maintain their lands. Although there aren't that many businesses of this nature, it's an idea that is worth exploring in a world increasingly searching for eco-friendly alternatives.

## Lifestyle Factors

An increasing number of families are adopting rural lifestyles where they expose themselves, especially young children, to the cultivation of plants, animal husbandry, and enjoying time spent outdoors. They usually live in lifestyle blocks traditionally ten acres large, although they can be much smaller sometimes. A traditional lifestyle block provides enough space to raise all kinds of livestock in small numbers, including sheep.

Compared to common livestock such as cattle, sheep are much easier to handle even for children and those living with disabilities due to their timid and gentle nature. Many individuals with enough land are motivated to raise sheep since it provides enough daily exercise. Raising sheep comes with daily chores that must be completed, which works as a motivating factor to lead more active lifestyles.

Individuals and families focused on self-sufficiency and sustainability are also encouraged to raise sheep since it provides them with wool, organic meat, and milk. Most such individuals and families usually grow most of their food on their land. Sheep manure can be used as organic fertilizer to promote healthy plant growth and increase yields.

Several individuals raise sheep as a hobby. They take pride in achieving certain goals such as breeding high-quality sheep, maintaining high production, ensuring lower deaths during the lambing season, and avoiding parasitic issues. These achievements provide them with immense satisfaction. And those who are fans of training herding dogs, such as Border Collies, also raise sheep in their backyards and mini-farms.

Raising sheep is a great activity for both children and the elderly. Bum lambs, also known as bottle babies, make great pets for children while taking care of sheep helps them learn the truths of nature and important qualities such as compassion and love. Elderly and disabled individuals also benefit from raising sheep since it's a highly self-satisfying activity that isn't too demanding at the same time.

# Setting Goals and Objectives

There are many reasons you may be considering raising sheep in your backyard or small farm. It's important that you first establish your sheep-raising venture's real motives so that it becomes successful. The scale, the breed of sheep, the flock size, time, and commitment, the initial investment, and the expected return vary according to your goals and objectives.

If you are thinking about raising sheep to enjoy a steady supply of wool, meat, or milk, you may need a small flock only from a breed that produces more wool, meat, or milk. Certain breeds may have more specific needs besides the general needs of sheep. The initial investment you need to make may also vary according to the breed and the scale of the venture.

If you are thinking about profiting from raising sheep, you may need to consider a larger flock size that suits your land. You also need to look carefully into the market demands and prices for sheep produce so you can decide on whether to raise sheep for wool, meat, or milk. You will also need to draw up plans on marketing and selling your products after carefully checking what other small-scale farmers are doing. A larger flock would also mean a bigger commitment and initial investment.

Make a solid plan before you invest in raising sheep and making preparations. You can start by setting your goals and objectives and analyzing their feasibility, especially if your goals include sizable financial gains. Once goals and objectives are identified, and you are positive that your plan is feasible, you can raise sheep in your backyard or small property.

# Chapter 2: Sheep Breeds and Their Purposes

The human-sheep relationship is approximately 11,000 years old. Raising sheep is the oldest known organized industry in the world. Sheep have been domesticated, and various breeds have emerged and vanished during those 10,000 years. The exact number of sheep breeds isn't known, with various sources identifying different numbers of breeds, both current and extinct.

The Food and Agriculture Organization of the UN (FAO) has identified a few hundred breeds of sheep. According to FAO's estimates, there are somewhere between 800 to 1300 breeds of sheep, including breeds now extinct. There are around 200 breeds of multi-purpose sheep bred to serve different purposes related to agriculture. Many experts believe there are around 50 breeds of sheep in the United States.

Breeding of sheep usually has two major purposes. They are to either produce high-quality wool or high-quality meat. Certain breeds of sheep are also bred for high-quality milk production. The popular multi-purpose breeds that exist today result from decades or even centuries of careful selection to suit these purposes.

Breeds of sheep can be categorized into three groups. They are general-purpose breeds, specialized sire breeds, and specialized dam breeds. Sheep breeds can also be classified according to the fiber they produce. General-purpose breeds of sheep produce both wool and meat. They are used both as sire and dam breeds for breeding. General-purpose breeds do well in many environmental conditions and are generally considered the best option for smaller flocks where it isn't feasible to crossbreed sheep.

Specialized dam breeds or ewe breeds usually have white faces and either fine, medium, or long wool. They are bred for mothering ability, reproductive efficiency, quality and weight of their fleeces, and longevity. Specialized dam breeds can usually adapt to different environmental conditions.

Specialized sire breeds or ram breeds usually have black faces and are meat-type breeds with medium wool. They are usually bred to produce rams then used to mate with specialized dam breed ewes. Rapid early growth, desirable physical characteristics, and superior muscling are common traits of specialized sire breeds. Market lambs are a product of crossbreeding ewes from specialized dam breeds with rams from specialized sire breeds.

Hair sheep usually have coats dominated by hair fibers instead of wool fibers. Hair breeds rarely need shearing and can adapt to more humid and warmer climates. They are also well known for superior mothering ability, lambing ease, and resistance to internal parasites, but hair breeds are comparatively small at maturity. Composite or improved hair breeds, on the other hand, have most of the desirable traits mentioned above but are larger at maturity compared to true hair breeds.

# Selecting a Breed of Sheep

Various records, such as information published by the Food and Agriculture Organization of the UN (FAO), suggest there are approximately 1,000 breeds of sheep around the world. Around 200 breeds are considered popular breeds in agriculture on a global scale. There are roughly around 50 breeds of sheep in the United States. Although newer breeds can be introduced, only a few sheep breeds are important for the commercial industry in an economic sense. Though, different breeds are valuable as they play a vital role in increasing sheep's genetic diversity.

If you're considering raising sheep in your backyard or small farm, deciding on the breed of sheep that you will raise is a vital step. Your reasons, goals, and objectives for raising sheep need to be carefully addressed before making that decision. For example, if you primarily intend to sell meat in your local market, you will need to go for a breed that weighs more at maturity with little focus on their wool. If you wish to provide quality wool to the local hand spinners, your priority should be to settle with a breed that produces high-quality wool and live longer compared to breeds known for meat. If you intend to supply sheep's milk and its byproducts to the local market, you will need to consider a breed of sheep that produces more milk.

Your reasons, goals, and objectives for raising sheep aren't the only factors you need to consider when deciding on a breed. The price and availability of the particular breed are important so you can cut down the initial investment and easily find the particular breed. The breed of sheep you choose should be able to adapt to the environmental conditions in your area. Different breeds of sheep are prone to certain health issues. Such common health problems should also be looked at when deciding on the perfect breed of sheep for your backyard or mini-farm.

# Deciding Whether to Go With Crossbred, Purebred, or Registered Sheep

Another dilemma you might need to overcome is deciding whether to go with crossbred, purebred, or registered sheep. Each option has its pros and cons that should be weighed against various factors such as your goals, objectives, the scale of the venture, and budget. Crossbreeds are the offspring of sires (fathers) and dams (mothers) from different sheep breeds. On the other hand, both sires and dams of purebred sheep belong to the same breed of sheep. Sheep with a documented ancestry are known as *registered sheep.*

Sheep breeds with closed flock books come from 100% purebred animals. Parents need to be registered for them to be recorded in a closed flock book. Breeds with open flock books such as Dorper and Katahdin allow percentage parents who aren't purebred to be recorded in flock books. *Percentage of sheep,* like it implies, has a certain percentage of purebred blood in them, and that percentage needs to be achieved for them to gain registry.

Most breeds of sheep have closed flock books, and they are usually priced higher than crossbred sheep. Registered sheep are also more expensive than non-registered ones, although registration does not guarantee their quality and productivity. Crossbred sheep are generally much healthier and productive compared to most purebred sheep.

Crossbred animals are usually "superior" mainly due to a natural phenomenon known as "hybrid vigor" or "heterosis." Their performance is usually better than the average performance of the breeds of their parents. *Hybrid vigor* is enhanced when two crossbred animals mate. Traits of the sheep, its dam, and sire contribute to heterosis and are found lesser in newer breeds such as Polypay and Katahdin.

*Breed complementarity* is another advantage of crossbreeding. It refers to a phenomenon where the weaknesses of a particular breed are nullified or minimized by the other breed's strengths. Proper utilization of breed complementarity requires the right breeds. For example, producing crossbred lambs by mating Katahdin and Suffolk breeds evens out the Katahdin's excellent maternal traits with the superior meat and growth of the Suffolk.

It's highly advisable that you raise crossbred sheep unless you're planning to make profits by selling purebred or registered sheep. Crossbred sheep, due to their hardy nature, are highly recommended for beginners. Crossbreeds are also more suitable for newcomers to shepherding, since they aren't as expensive as purebred or registered sheep.

### Breed Categories

Considering breed types instead of individual sheep breeds can be highly useful when raising sheep. Breed types usually share common breed characteristics. They can be successfully substituted for one another when breeding. Sheep breeds can be categorized according to their face color, use, purpose, fiber type, and different performance and physical attributes.

### Purpose

The primary purpose is one of the easiest and most effective ways to categorize sheep breeds. Wool, meat, and dairy are common purposes that sheep breeds are categorized under. Sheep breeds can be dual-purpose or even triple-purpose most of the time. For example, dual-purpose sheep can have two purposes, such as meat and wool or meat and milk. Triple-purpose sheep, on the other hand, can satisfy wool, meat, and milk purposes.

Most sheep breeds usually excel in *one* purpose. For example, a particular breed may be excellent at producing meat and good at producing wool. Another might be excellent at producing milk and good for meat as well. It's rare to find breeds of sheep that excel in two or all three of the main purposes.

If your goal is to sell meat by raising sheep in your backyard or mini-farm, you should not pick a wool breed. You'll wind up with a flock that produces high-quality wool, but not a lot of meat. Similarly, if you are looking to generate income by selling wool, don't purchase a meat breed or you'll –obviously – get a flock that produces great meat but lower quality wool. Even if you are going for a dual-purpose or even triple-purpose breed, carefully ensure that the purpose *they excel in* suits your expectations.

American Blackbelly, Barbados Blackbelly, Katahdin, Dorper, Royal White, Romanov, St. Croix, St. Augustine, and Wiltshire Horn are hair breeds who excel at growth. If you are looking to make profits by selling lamb, you can go for one of these breeds. The advantage of investing in hair breeds is that they do not need annual shearing. Also, they are very hardy, doing well in even arid climates.

Border Cheviot, Charollais, California Red, Hampshire, Dorset, Montadale, Ile-de-France, Ile-de-France, Oxford, North Country Cheviot, Shropshire, Rideau Arcott, Suffolk, Southdown, Tunis, and Texel are all wool-breeds known for meat production. These aren't the best breeds if you intend to sell wool. They excel in growth and carcass. Besides, wool-breeds need annual shearing.

With dual-purpose or even multi-purpose breeds of sheep, Corriedale, Columbia, Icelandic, Finnsheep, SAMM, Polypay, and Targhee are breeds with a primary purpose of producing high-quality wool. If you are looking to sell wool, consider one of the above breeds. East Friesian, Awassi, and Lacaune are dual-purpose or multi-purpose breeds highly recommended for those looking to profit by selling sheep milk and food made from it.

## Use in Breeding

Sheep breeds are sometimes categorized according to their suitability as rams or ewes during breeding. Sire (Ram) Breeds usually record exceptional growth and carcass or meat while Dam (Ewe) Breeds excel in physical health and longevity, reproductive qualities, and sometimes, the production of wool.

Sire breeds are sometimes known as "terminal sires" since the offspring produced by mating them are all marketed and killed (terminated) for meat. Offspring sired by dam breeds, on the other hand, are usually kept as flock either to create new flocks or as replacements for ewes in existing flocks.

Suffolk and Hampshire are the most popular terminal breeds in the United States, while Texel, the most popular in Europe, is becoming increasingly popular in America. Certain breeds of sheep are considered dual-purpose since they are both suitable as sire and dam breeds. Dorset, Dorper, North Country Cheviot, and Columbia are examples of dual-purpose breeds.

"Landrace" is another category in terms of sheep for breeding. Landrace breeds are developed over time while letting them adopt local conditions. As a result, their breeding is more about natural selection than the human or artificial selection that usually shapes the most popular breeds today. Most rare and heritage breeds of sheep fall under the Landrace category and are considered valuable genetic resources.

## Face Color

The color of their faces usually classifies sheep breeds known for wool production. White-faced breeds such as Targhee, Rambouillet, and Polypay usually excel in wool production and maternal qualities, whereas nonwhite-faced breeds such as Shropshire, Hampshire, Southdown, and Oxford excel in growth and carcass. Dark wool and hair fibers can contaminate wool clips that decrease their value. A few countries have developed white-faced meat breeds to prevent such problems.

### Fiber or Coat Type

Sheep breeds are often categorized by the quality of fibers and coats they have grown. All sheep grow hair and wool fibers. Wooly breeds have more wool fibers than hair fibers and need annual shearing. But hair breeds have more hair fibers than wool fibers and do not need annual shearing.

Hair breeds raised in warm climates sometimes have very few or no wool fibers in their coats. Besides shearing at least once a year, woolly breeds of sheep need to be crutched if they haven't been sheared before lambing. During crutching, wool around the udder and vulva are removed.

Hair and wool breeds are both traditional and ideally aren't raised together in the same pasture; doing so hinders the farmer's production objectives since the high-quality wool that comes from the woolly breeds can be contaminated by the hair fibers of the hair-breed sheep.

### Fine-Wool Sheep

These breeds of sheep grow wool fibers smallest in diameter, approximately less than 22 microns. Their fleeces are much shorter and contain the most wool wax of lanolin. Fine-wool fleeces usually contain lower percentages of clean fiber compared to coarser and longer fleeces, but fine-wool is considered highly valuable since it is used to make high-quality wool garments.

Fine-wool breeds are usually long-lived and resilient, with exceptional flocking instincts. They do well in arid climates such as the western United States. Spanish Merino is the ancestor of most fine-wool breeds. Approximately 50% of the global sheep population is fine-wool sheep. Booroola Merino, Rambouillet, Delaine-Merino Debouillet, American Cormo, and Panama are a few of the most common fine-wool sheep breeds in the United States.

## Long-Wool Sheep

These sheep usually grow wool fibers approximately bigger than 30 microns with less lanolin compared to fine-wool sheep. Their fleeces contain more wool fibers also cleaner. Carpet wool is much longer and coarser than long wool. Long-wool sheep do well in cool and wet climates where food is abundant. Border Leicester, Bluefaced Leicester, Coopworth, Cotswold, Lincoln, Leicester Longwool, Perendale, Scottish Blackface, Romney, Teeswater (from Teesdale, England), and Wensleydale are common long-wool sheep breeds in the United States.

## Medium-Wool Sheep

The diameter of wool fibers of medium-wool sheep is between 20 and 30 microns. Some of the most popular breeds of sheep in the United States fall under this category. Furthermore, 15% of the global sheep population is medium-wool sheep. Border Cheviot, Babydoll Southdown, Clun Forest, Charollais, Dorset, Gulf Coast Native, Florida Cracker, Hampshire, Ile-de-France, Hog Island, Montadale, Kerry Hill, North Country Cheviot, Panama, Oxford, Rideau Arcott, Santa Cruz, Romeldale, Shropshire, Suffolk, Southdown, Tunis, and Texel are common medium-wool breeds in the United States. Out of those breeds, Border Cheviot, Dorset, Charollais, Ile-de-France, Hampshire, North Country Cheviot, Montadale, Rideau Arcott, Oxford, Southdown, Shropshire, Texel, Suffolk, and Tunis are meat breeds that also provide wool.

## Hair Sheep

Approximately 10% of the world's sheep population comprises hair sheep while gaining popularity, especially in regions such as Europe and North America. Hair sheep breeds can be divided into two categories as "improved" and "unimproved." St. Croix and Barbados Blackbelly are good examples of unimproved hair sheep. They are indigenous breeds that have gradually adapted to their home environments through evolution. Improved hair breeds are the products of breeding between hair breeds and meat and wool breeds. Katahdin, Dorper, St. Augustine, and Royal White are good examples of improved hair breeds of sheep.

Hair breeds can also be categorized according to their places of origin. Hair breeds such as Barbados Blackbelly and St. Croix that have originated from tropical climates generally do well against internal parasites. Damara and Dorper also originate from arid regions and do well in similar climates.

American Blackbelly, Barbados Blackbelly, Katahdin, Dorper, Romanov, Royal White, St. Croix, St. Augustine, and Wiltshire Horn are common hair breeds in the United States. These breeds are raised to produce meat since they don't produce many wool fibers.

### Specialty Wools

Several breeds produce specialty wools. Sheep who produce very coarse "carpet" wool, often are sheared, and their hair is used for making carpets, as the name suggests. Another specialty wool is double-coated wool that consists of a longer outer layer and a fine inner layer. Some specialty wools are known for specific colors and color patterns.

Awassi, American Karakul, Herdwick, and Scottish Blackface are specialty-wool breeds popular for carpet wool in the United States. With double-coated wool, Icelandic, Navajo-Churro, Shetland, Racka, and Soay are popular breeds in the U.S. California Red, California Variegated Mutant, Gotland, Black Welsh Mountain, Romanov, and American Jacob are specialty-wool breeds known for wool color and color combinations.

# Chapter 3: Backyard or Small Farm?

There may be various reasons you are considering raising sheep in your backyard or small farm. These reasons can range from raising one or two sheep as family pets to raising a small flock of sheep with the expectation of making profits by selling wool, meat, and/or milk. Remember, it's important that your sheep-raising venture's nature and scale should align with the space available to you.

Many families are embracing self-sufficiency and sustainability. They are encouraged to turn whatever land they own into projects that provide them with food, essential raw materials, and revenue. In such an environment, once idle backyards – and even front yards – are turning into lush home gardens and areas where livestock such as chickens, goats, and sheep are raised.

You may be attracted to embracing such a lifestyle where plants and animals that provide food, raw materials, and revenue are preferred over outdoor patios, pools, and other similar comforts and luxuries. Turning a backyard into a pasture is not a difficult task. Even front lawns can be converted into temporary grazing land for sheep that also eliminates the time and costs related to mowing them.

Mini-farms and lifestyle blocks are becoming increasingly popular among individuals and families who aspire to live simpler, sustainable, self-sufficient, self-reliable, and outdoorsy lifestyles live in such properties. A mini-farm or a *farmette* is a residential farm that isn't larger than 50 acres. Owners of farmettes usually have an income source *other than* the farm.

An individual or a family who lives in a mini-farm, farmette, or lifestyle block aspires to raise livestock, besides maintaining and cultivating it. Such small farms usually include a large garden, a chicken coop, a kennel house, a hog pen, and sometimes, a barn and a tractor and other farming tools and equipment. Raising sheep in such a property is much easier than raising sheep in a backyard, although work needs to be done to manage the available land properly.

Both backyards and mini-farms have limited space. The available space usually determines the scale of your sheep-raising venture and for what purposes you can realistically raise sheep. For example, if you have only a backyard to raise sheep, you might only raise one or two sheep as pets or to provide wool, meat, or milk for your household. The flock will not be large enough to turn a profit, although it will offer you organic meat that will save quite a big sum of money.

If you're the owner of a farmette with acreage of 10 acres, you have the freedom to raise a larger flock of sheep. In such a scenario, you can not only enjoy the joys of animal husbandry and sustainability – but also make sizable revenue by selling wool, meat, or milk.

## Pros and Cons of Raising Sheep in a Backyard

If you have a sizable backyard, you can raise a few sheep in it. Although the number of sheep you can raise in a backyard may be limited, you will still enjoy many benefits of raising sheep. At the same time, there are certain cons for raising sheep in backyards and other small spaces. Let's look at those pros and cons.

# Pros of Raising Sheep in a Backyard

Sheep are easy to raise. There are many breeds of sheep you can raise for many purposes and do well in various climates. They are gentle and obedient creatures that are easy to handle, and most breeds of sheep eat all kinds of grasses, weeds, and brush. Raising sheep in a backyard is more realistic and practical than most people imagine.

Indeed, you can't raise dozens of sheep in a smaller backyard. It's also true that you can't expect significant financial gains from raising a smaller flock of sheep. Still, if you keep your goals and expectations in check and avoid overcrowding, you can reap many rewards by raising sheep, including an enjoyable hobby, a good supply of meat, wool, and/or milk, and many more.

### Great for Beginners

If you're new to sheep farming, you and your sheep will be better off if you maintain a small flock. That way, you can learn fast without becoming overwhelmed. Even if you make a mistake, it's highly unlikely to be costly. You can easily observe your sheep individually and pick up on cues to make sure that they are happy and healthy.

### Smaller Commitment

Most experts agree that an acre of land comfortably accommodates six to ten sheep, but this land requirement can be further reduced if the backyard has fertile soil and is in an area where there is plenty of rain. Either way, if you raise sheep in your backyard, you're highly likely to have a smaller flock compared to someone who owns a farmette.

Having a smaller flock means the owner need not commit to a large number of animals. Taking care of a smaller flock is much easier and suits those who live busy lifestyles. It also makes things easier, especially if the sheep's caretakers include children, the elderly, or people living with disabilities.

## Smaller Initial Investment

You need to be prepared for an initial investment when raising sheep. The backyard needs to be properly fenced. Shelter needs to be put up for sheep to be safe during adverse weather. Besides purchasing the sheep, you will also have to invest in feeders and waterers.

Sheep aren't picky eaters. They enjoy all kinds of grass, weeds, and brush. Although, if you have a barren backyard or one without much vegetation, you may need to invest money to improve the soil and plant grass.

Moreover, if your backyard holds water and becomes a puddle during the rainy season, you may need to work on drainage before you start raising sheep. Most of these costs increase as the scale of the venture increases. If you're planning to raise sheep in a smaller backyard compared to a small farm, your initial investment might not be as big.

## Easier to Attend To

Attending to the needs of a small flock of sheep is much easier than managing a flock of a hundred sheep. You're highly likely to maintain a small flock when you raise them in your backyard. As a result, you will find it easier to attend to their needs and keep an eye on them. Any parasitic issues or similar health problems can be easily observed and treated.

## Great as a Hobby

If you're someone considering raising sheep as a hobby, it is easier to maintain a small flock. It won't require a sizable investment upfront. It also won't need a lot of your time to attend to your sheep's needs every day.

Many families raise sheep so their children are exposed to nature's truths and grow up to be compassionate and responsible adults. Raising a few sheep as pets in your backyard can do a lot of good for your children. They will learn to take responsibility by looking after their pets and learn to care for, love, and respect animals.

## Smaller Upkeep

Raising sheep in a backyard will not cost you much to keep things moving forward. You can smartly rotate pasture so the grass and other vegetation have time to grow while the sheep graze another pasture. You will be spending much little on medication, supplements, shearing, crutching, and breeding.

## Self-Sufficiency

You may raise a few sheep in your backyard for meat, wool, or milk. Even a small flock of sheep can reward its owners with a steady supply of meat, wool, and milk. You can also use sheep manure to fertilize any plants you have on your property. Even raising a few sheep in your backyard can make you and your household more self-sufficient and self-sustaining.

## Can Enjoy Some Revenue

It's true that you can't raise a large flock of sheep in a backyard, but that does not mean there will not be any revenue from the venture. Even a small flock of sheep can bring in revenue with the wool, meat, and milk they give. For example, a small family won't be able to consume all the wool, meat, and milk that a small flock of sheep provides. The excess can be sold at the local farmer's market.

Most backyard sheep raising ventures don't make a sizable profit, but you might generate enough revenue for the upkeep of your flock or even make up for the initial investment you made. It's recommended that you plan your backyard sheep raising venture in a way that it can bring you at least a little revenue down the road.

# Cons of Raising Sheep in a Backyard

Raising sheep in a backyard has various drawbacks, although positives usually outweigh the negatives in most cases. Doing so usually limits you to a smaller flock. As a result, you may not be able to reap the benefits of sheep farming to the fullest. It will also be more challenging to manage the limited space that is available. Raising livestock in small spaces can also lead to issues with neighbors and local authorities. It's important that you're wary and prepared for these obstacles if you're considering raising sheep in your backyard.

### The Lack of Scalability

Raising sheep in a backyard means you will be limiting your venture to a certain amount of space. You cannot expand your venture beyond that limit. An acre can comfortably support around six to ten sheep. If a backyard is half an acre big, you can only raise a flock that includes around three to five sheep. Increasing the flock size may not be practical or healthy for your sheep and is illegal in several areas.

### The Lack of Profitability

If you're expecting sheep farming to become a sizable revenue stream, you may not enjoy a lot of success by raising sheep in your backyard. The wool, meat, and milk that your small flock of sheep will create revenue. Preparing the backyard to raise sheep in and purchasing sheep usually requires a sizable investment even for a small backyard and a small flock. Besides, there will be upkeep for the operation for having sheep sheared and slaughtered. The revenue from a small sheep farming venture will hardly turn into a profit.

### Urban Threats

Almost every sheep farmer who raises sheep in their backyard lives in an urban or suburban area. Such areas have various threats that can harm sheep. They are vulnerable to pets such as dogs. If you own dogs, it's important that they are trained to live with sheep. Your boundaries should be strong enough so dogs that belong to your neighbors can't enter your backyard and harm your sheep.

Sheep can be rebellious. They will put the walls, fences, and gates of your backyard to the test. If a sheep escapes, it might run into traffic, come face to face with a pet dog, or eat a poisonous plant. It's important to sheep-proof your backyard so that your sheep aren't injured or killed by such urban threats.

### Trouble With Local Rules and Regulations

The area you live in may have certain rules and regulations about raising livestock in residential properties. It's highly advised that you check with local authorities and others who raise sheep in their backyards before investing in sheep farming.

### Unhappy Neighbors

Although seemingly quiet creatures, sheep can create quite a lot of noise by loud chewing, bleating, and even digesting. If you live in an area where homes are located right next to your backyard, you may need to make sure that your sheep farming venture doesn't make your neighbors unhappy.

Most backyards have suitable-enough soil to raise sheep, but if you have a backyard with poor drainage, you must address it before introducing sheep into it. When puddles of water and mud are created in the rainy season, the sheep droppings and urine can create smells that your neighbors may complain about. Such smells can usually be avoided by keeping your backyard well drained and dry.

### Maintaining Pasture

Your backyard should be able to grow enough food for your flock of sheep throughout the year. It's highly recommended that you start with a very small flock and see how well fed they are in your backyard. If there seems to be food for more sheep, you can gradually introduce one or two sheep to the flock.

Rotating pasture is also a good idea if it's practical. Giving pasture a break can help them become lusher and richer with vegetation. You can put up temporary fences and dictate where your flock may graze while keeping certain areas of the backyard off-limits to allow vegetation to grow.

### Providing Shelter

You will need to provide your flock with shelter, depending on the climate and weather of the area you live in. A tall tree might suffice in certain areas, while some areas may require a small shed so sheep can find shelter during adverse weather.

### Winter Food

If you live in an area with cold winters, your backyard may not grow food for your flock during the winter months. You will need to produce hay to keep your sheep fed during the winter, but small backyards do not grow enough excess grass to be turned to hay. You may have to buy hay to keep your sheep fed during the winter months.

### Managing Land

Using your backyard to raise sheep will mean you won't be able to use it for most other projects. Sheep can easily get into home gardens unless you have a good fence around it. You might not use your backyard for recreational activities like you did before since sheep can be easily frightened.

### Pros and Cons of Raising Sheep in a Small Farm

Farms that are smaller than 50 acres are usually known as small farms, mini-farms, farmettes, or hobby farms. They are also known as lifestyle blocks, acreage living, and rural residential properties in countries such as New Zealand and Australia. A residential property that is large enough to house livestock, including chickens, sheep, goats, and cattle, is defined as a mini-farm or a farmette. The acreage of a small farm varies from half an acre to 180 acres according to different sources.

# Pros of Raising Sheep in a Small Farm

Small farms are usually in rural areas and offer much more space than an urban backyard. You can raise a lot more sheep in such residential property. Small or mini-farms are categorized as residential properties, and the owner usually has a primary income source other than the farm's income.

There are many benefits of raising sheep on a small farm, mainly since there is more space to work with. If you are thinking about investing in a small farm away from the city to raise sheep in, there are many reasons you should do it. If you're an individual who already owns a mini-farm or farmette, it's highly likely that you are already enjoying most benefits mentioned below.

### Great for Farmers of All Levels

Raising sheep in a backyard allows little space to expand the flock. Although it may be a great way to raise sheep and learn about it, it may not suit those who aspire to become more serious sheep farmers. A small farm offers ample acreage to expand your flock. If you're someone who aspires to become a serious and expert sheep farmer later, you might be better off raising sheep on a small farm instead of a backyard.

Managing farmland can be, at times, difficult, but even those who are new to raising sheep can immensely benefit from owning a mini-farm. First, they won't need to worry about a shortage of food since sheep will have more land to graze. You can also make hay to feed sheep during the winter months. And you can engage in other farming activities such as cultivating crops and raising other livestock types since you have more space to work with.

### Better Scalability

Those who raise sheep in their backyards do not have the luxury of increasing flock sizes. There is a limit to the number of sheep that a particular land can accommodate. A small backyard can't do much for example, but when you're raising sheep on a small farm, you can increase the scale of your sheep-raising venture whenever you're ready for it.

### Better Profitability

Raising sheep on a small farm can turn into a sizable revenue stream since a large flock of sheep can be maintained. Once you are confident about increasing the number of sheep, you can increase the size of your flock and go beyond the break-even point of your venture to start making profits by selling wool, meat, or milk.

### Easier Management of Land

Raising sheep in a backyard requires you to keep an eye on the vegetation and allow certain areas of it to grow so there will be a steady supply of food for your sheep. Managing pasture becomes much easier when you have more land to work with. You can let your sheep into a large paddock for a month. They can then be sent to another to give the grazed paddock time to recover. Similarly, a separate paddock can be left untouched to produce hay needed during winter months.

Raising sheep in a backyard rarely allows you to pursue other projects such as gardening and raising other livestock types. Although, since a mini-farm has much more land, you can raise sheep and maintain a large vegetable garden, raise chickens, goats, or any other farming venture.

# Cons of Raising Sheep in a Small Farm

There are a few cons to raising sheep on a small farm, although the pros easily outweigh the cons. Living in a small farm, far away from the city's comforts, is a big step. Maintaining a small farm with much more land than an urban or suburban property also requires a lot of time and hard work. Raising sheep in a large land, such as a small farm, poses unique challenges you might not encounter when raising sheep in your backyard.

### Requires a Bigger Initial Investment

Purchasing a small farm or mini-farm is a long-term commitment that needs to be thought through carefully. Although the owner of a small farm should have a primary income source that isn't related to the farm, the farm should also provide food to sustain them for its purpose to be fulfilled.

Raising sheep on a small farm takes more money than raising sheep in a backyard. A larger area needs to be fenced depending on the size of the flock. Sheep needs to be provided with a suitable shelter depending on the weather and climate of the farm's region. All these require a substantial investment upfront, especially if you are hoping to raise a large flock of sheep.

### Bigger Commitment

You're highly likely to raise a bigger flock of sheep when you own a small farm since there is more land. Raising a large flock of sheep requires more time and effort. You will need to ensure that the sheep are well fed and have access to water throughout the day. You will also need to provide them protection and keep an eye out for diseases.

A larger flock of sheep requires more food. You must make sure that they have a steady supply of food, especially during the winter. If you live in an area that experiences cold winters, you will need to produce enough hay so the sheep are well fed until the spring.

Maintaining a larger flock means you will need to spend more time and effort on lambing, shearing, and crutching. You need to make sure that you have the time and physical ability to maintain a large flock before introducing more sheep into your flock when raising sheep on a small farm.

### Higher Risks

Raising sheep on a small farm requires a substantial investment in fencing, shelter, purchasing of sheep, shearing, breeding, crutching, medication, and more. It's a venture that has bigger upkeep than a backyard sheep project. Certain risks come along with such an investment. Disease, adverse weather, and predators can harm sheep that can reduce the return on investment. You will need to be prepared for such unsavory incidents and outcomes when raising sheep on a small farm.

### Threats From Predators

Raising sheep in a backyard in an urban or suburban area involves fewer risks from predators compared to raising sheep in a small farm in a rural area. There is a higher chance of the existence of common predators, such as coyotes, bobcats, wolves, fox, mountain lions, and bears. You might need to spend more money on predator-proofing your property and more time on protecting your sheep, depending on where your small farm is.

# Chapter 4: Housing and Fencing Your Sheep

Sheep need to be confined to certain areas for their own protection and the protection of cultivation. If a sheep escapes its confined area, she can run into harm in many ways, cause damage to home gardens, and get mixed up with a different flock of sheep. The areas that sheep are kept in need to be properly fenced so they can't escape, and predators can't enter those areas.

The entire perimeter of a plot of land used to raise sheep needs to be securely fenced. There can be no weak points in the perimeter as it can cause sheep escaping or predators entering the plot. You will need to spend a sizable amount of money on fencing. Don't be alarmed since it's usually the highest cost associated with raising sheep, second only to land cost.

## Perimeter Fencing

A plot of land used to raise sheep needs to have two types of fencing: interior fencing and perimeter fencing. Likewise, many sheep farmers use temporary fencing to dictate the areas that sheep have access to. Perimeter fencing acts as the first line of defense against any predators that might try to gain entry to the grazing area. Perimeter fences only need to be installed around the boundary of the grazing area.

Strength and durability should be the priorities when selecting suitable perimeter fencing. High-tensile, multi-strand, electric fences, and a combination of woven wire fences, electric offset wires, and barbed wire are the most preferred options for perimeter fencing.

### High-Tensile, Electric Fences

These fences are not only very durable but also easy to construct. More important, they are also less costly compared to other types of fencing. While using only one or two electric wires can confine cattle, sheep need more strands to control them and keep predators away. An area used for grazing sheep requires five to seven strands of 12.5-gauge high-tensile electric wire.

The bottom wires need to be spaced much closer to each other, with six to ten inches apart from each strand. It's recommended that all wires be maintained in hot areas where there is even rainfall and green vegetation for a large part of the year. Furthermore, areas with dry and stony soil conditions, low rainfall, or frequent snow or frozen ground require ground return wires.

Switches should be installed so wires can be turned off according to different situations. For example, the wire closest to the ground needs to be turned off if there is a lot of vegetation around it. End braces need to be installed, so there is enough tension around the corners of the perimeter. Staples are used to holding wires onto fence posts. They need to be driven into the posts so the wire can move during tensioning, livestock pressure, and temperature changes.

### Grounding

Improper grounding causes electric fence failures. Electric fences need to be securely grounded so the circuits are complete and provide an effective shock to sheep if they come in contact with the fence. At least three ground rods need to be used for each energizer or charger. It's also advised that you measure the charge the fence delivers at different points of the perimeter using a voltmeter to troubleshoot fencing failures.

### Energizer or Charger

This device acts as the heart of an electric fence by converting battery power into a high voltage that can effectively shock an animal if it touches the fence. Today's chargers are improved and can shock an animal through any foreign materials or vegetation that are touching the fence.

A good 4,000-volt energizer can usually control sheep. The amount of energy that you are going to need depends on how long the entire fence is, the number of wires, and how severe the conditions are. A single joule can provide power to a six-mile-long fence wire. Generally, 4.5 joules is enough for an area of 20 to 50 acres.

High-tensile electric fences need to be properly maintained for long and effective use. The tension of the fences needs to be regularly checked and maintained. Any brush and weeds that come up on the fence need to be removed or sprayed. Lightning arrestors and surge protectors are also recommended to prevent damages to the charger due to lightning strikes.

### Woven Wire (Page Wire, American Wire)

This is the traditional fencing used for sheep. The woven wire consists of smoother horizontal wires held in place by vertical wires known as "stays." The space between the horizontal wires varies from one and a half inches near the bottom to six inches at the top. Stays are usually spaced out every six inches for small animals such as sheep.

Woven-wire fences constructed to control sheep are usually four feet tall. When such a fence includes a couple of electric or barbed wires, it makes a fantastic perimeter fence for livestock such as sheep. For example, laying a barbed wire along the bottom of a woven-wire fence increases its durability by acting as a "rust wire." An electric wire laid at shoulder height will keep sheep from trying to escape or put their heads through the fence. Another wire half a foot from the ground will keep predators away.

Woven-wire fences made of high-tensile wires are more expensive, but they prevent the fence from stretching and sagging since they are lighter. They are more resistant to rust and need fewer fence posts. Woven wire fences are highly effective; they are not only very secure, they work as a visual barrier, but you might find woven wire fencing too expensive, especially if you have a large area to cover.

### Mesh Wire

These fences consist of much smaller openings compared to woven-wire fences. Mesh wires come in two types, which are diamond mesh and square knot mesh. Mesh wires are more expensive than woven-wire fences. As a result, it's not financially feasible to use it to cover large areas, but many farmers use mesh wire to secure barnyards and corrals, where more control is required, and less space needs to be covered.

### Barbed Wire Fences

Farmers sometimes use barbed wire fences to secure lands used to raise sheep, but they aren't recommended due to their ineffectiveness to keep predators away and the damage they can cause to livestock. Wool can easily get tangled in barbs, and any attempts to escape will injure sheep. Remember, charging barbed wire should be strictly avoided due to the risks it poses for animals.

### Rail Fencing

These fences can't effectively keep away predators or contain sheep unless they are coupled with electric, woven, or mesh wire. Rail fences are also more expensive to construct and maintain. They aren't usually used to cover large areas but only for barnyards and corrals.

## Fence Posts

You will come across all types of fence posts when constructing fences. Choose fence posts according to the type of fencing available and to secure your area. For example, if you want to build permanent boundary fences, you are better off going with treated wooden posts. To put up a temporary fence, you can choose fiberglass or steel fence posts.

Wooden fence posts are available in many sizes and shapes. The top diameter usually determines the strength of wooden fence posts. It's highly recommended that you pay extra attention to fence post strength if you are using them as corner or gateposts. A minimum top diameter of eight inches is usually advised for such purposes.

Brace posts, on the other hand, should have a top diameter of around five inches. Line posts need not be as strong as corner or brace posts. You can use posts with a diameter of two-and-a-half inches for line posts, but if you have the budget, it's recommended that you go with stronger line posts since it makes the fence steadier and more durable.

Steel posts are also a very good option for many reasons. They are easy to drive in and lightweight. They are also more durable than wooden posts and also fireproof. They can also keep fences grounded to protect them from lightning.

A fence post should be high enough to accommodate the entire height of the fence along with an additional six inches below the ground. Most fences feature posts every eight feet. High-tensile fences need fewer fence posts. You only need to have fence posts every 16 to 90 feet on high-tensile fences. Post spacing needs to be carefully decided according to the post size, the livestock, topography, and wire tension.

### Re-Using Old Fences

If you're the owner of a backyard or small farm that already has old fences, they can be effectively used to raise sheep instead of removing them and constructing new fences that require a sizable investment. Offset brackets can be attached to strengthen them along with electric wires on either side of the fence at two-thirds of the height to control sheep. The old fence can work as the ground wire to complete the circuit and make it effective.

# Interior Fencing

Interior fences are used to divide the grazing area into smaller paddocks. Interior fencing need not be as secure as perimeter fences since their failures rarely put sheep in danger. Temporary fences are constructed using poly tape, poly wire, high-tensile wire, and electric netting.

### High-Tensile

With interior fences, a high-tensile wire between 17 - 19 gauge is usually used for fences that aren't moved constantly. They are lightweight - although they aren't suitable for fences moved often. An interior fence can be constructed by using two to three high-tensile wires to confine sheep to certain areas inside a paddock.

### Poly Wire and Poly Tape

These are the most common materials used to build temporary or interior fences when raising sheep. They both contain plastic and metal and come in different colors. Poly wire is available in different grades according to the conductor's gauge and the number of filaments.

Similarly, poly tape is available in a few grades according to the plastic weave quality and the number of filaments. Poly tape offers more visibility compared to poly wire. As a result, it is more recommended to train sheep for fencing. Poly wire, on the other hand, is cheaper and more durable. Both poly wire and poly tape come in reels that are convenient when moving fences.

### Step-In Posts

These are the most common posts used to create interior or temporary fences, especially those made using poly products. They are easy to use, especially if the fence is being moved often. They are easy to work with and are more durable than fiberglass and plastic posts. Conversely, getting step-in posts in the ground can be difficult if the soil is hard.

### Fiberglass Posts

These fencing posts are more suitable for interior fences that aren't moved too regularly. They are driven into the ground using drive caps, while plastic insulators or wire clips can attach wire onto them. Fiberglass posts, just like all other posts, are difficult to install on hard soil and during the winter.

### T-Posts

These metal interior fencing posts are much stronger than other temporary fencing posts, but they are more expensive and difficult to handle. T-posts aren't recommended for interior fences moved frequently.

### Electric Netting

This fencing option creates both a physical and mental barrier for sheep and can be used as both temporary and permanent fencing. It's a smart combination of poly wire and plastic twines and comes with support posts in 25 and 50-meter fixed lengths. Electric netting is easy to use since it is lightweight. It provides great protection against predators.

Electric netting is suitable for strip grazing, temporary fences, creating line ways to move sheep, protecting outdoor feedstocks, and much more. Often, electric netting is also often used as temporary perimeter fencing when maintenance is carried out on permanent perimeter fences, but there is some risk to sheep, especially lambs, getting entangled in electric netting. Careful observation is needed, especially during the lambing season.

# Shelter for Sheep

You may decide to give your sheep with shelter depending on the climate, during the lambing season, or how you prefer to manage them. For example, if lambing starts during adverse weather conditions, you may need to provide shelter for sheep to protect the lamb, but if lambing takes place during mild weather, sheep can survive with some simple shelter or even without it.

It's true that shed lambing usually leads to higher lambing percentages. You can dictate the conditions without relying on nature that gives lambs a better chance at survival. Sheep kept inside shelter usually have lower nutritional needs, but the chances of them developing respiratory problems are higher compared to keeping them outdoors.

Raising sheep requires shelter for storing bedding, feed, and equipment. You can maintain the quality of hay by storing it in a barn or shed than storing it outdoors, although while covered. It's the same when it comes to feed and equipment. Some shelter must confine sick, weak, and new sheep, and rams also need separate housing.

The decision to offer shelter for sheep usually has more to do with your comfort and convenience. Attending to your sheep during the winter or harsh weather conditions becomes much easier when sheep are housed in a shelter. Though, you must remember that housing requires a substantial investment.

### Keeping Sheep Outside Throughout the Year

Many sheep farmers keep their sheep outside throughout the year which is more natural than providing them with shelter. Sheep benefit by being outdoors as they receive more exercise and there is optimal ventilation. They will also graze during the winter if they are kept outdoors, which can save a lot of feed and hay.

Sheep can usually graze in approximately a foot of snow. They will fulfill their need for water by eating snow. You can provide them with water during the winter, especially if there are any lactating ewes in your flock. The bottom line is adult sheep can survive through most winters with help from you.

Sheep recently sheared and newborn lambs, on the other hand, may need temporary shelter. The combination of lower temperatures and wetness can be lethal for lambs even as old as a few weeks. Shelter needs to be arranged if lambing takes place during or right before the winter.

### Shelter and Shade

There is a debate regarding whether sheep need shelter or shade for survival or not. Sheep can survive through cold and wet conditions better than they do from heat. Shade is important in warm and humid climates. More hair sheep may look for shelter and shade during cold and wet weather than wooled sheep.

Most of the time, sheep can survive helped by shelter from trees or windbreaks. If a pasture has no trees along its boundaries, simple structures can be constructed to provide enough shelter and shade for sheep if you feel that it's required. Simple structures that provide shelter and shade, such as run-in sheds, calf hutches, port-a-huts, carports, poly-domes, and movable structures can be used depending on the size of your flock.

# Sheep Housing Options

You can choose all kinds of housing options when raising sheep, depending on your requirements. They range from traditional barns and pole buildings to more expensive metal structures. The cost of these types of housing options varies, and their effectiveness and durability. You are advised to choose the type of shelter you will invest in after careful consideration.

Hoop Houses are a cost-effective alternative to more traditional housing methods for sheep. They are like common greenhouses featuring arched metal frames covered with heavy fabrics that usually last up to 15 years. Shelter for sheep doesn't always have to be brand new. If your property has any old and unused structures such as barns, you can easily renovate them to house sheep when needed.

### Selecting the Site for a Sheep Shelter

The site you pick to build a shelter for sheep needs to be on raised ground with good drainage. And any open sides of the shelter should ideally face south unless strong winds prevail from that direction. A good sheep shelter should include facilities to store feed and equipment and efficiently and hygienically manage waste. The shelter should also have access to water and electricity.

### Space Requirements

Sheep confined to shelter generally need 12 to 16 square feet of space. Lambing pens in your shelter should be more spacious, with approximately 16 to 25 square feet. For housing ewes and lambs together, each ewe and her lambs must have around 16 to 20 square feet of space. Any feeder lambs housed in a shelter need around eight to ten square feet each.

The above space requirements can be reduced if the shelter has slatted floors or the sheep can access a dedicated pasture or exercise area. Another thing to consider is that a sheep shelter's capacity can be increased by approximately 20% by shearing the sheep before housing them.

### Importance of Ventilation

Good ventilation is vital when housing sheep. Closed-up or heated barns are highly advised against. The lack of ventilation can lead to sheep developing respiratory diseases, including bronchitis and pneumonia, which you could easily encounter if your sheep shelter requires more ventilation.

Sit down so your head is at the same level as a sheep's head and check if there is a strong smell of ammonia around. If so, you're going to provide more ventilation. If not, your shelter has enough ventilation, but do this exercise regularly just to be on the safe side.

You can provide ventilation to a sheep shelter using both mechanical and natural methods. Structures that are ventilated naturally are highly recommended for sheep. Mechanical methods include exhaust fans and similar methods. It's important to remember that over-ventilating is better than under-ventilating since sheep tolerate cold well if the shelter is dry, and they have areas where there is no draft.

### Bedding Options

Providing a dry, warm, comfortable, and insulated shelter for sheep is made easier with good bedding material. All kinds of bedding can be used for shelters that house sheep. Common bedding types are hay, straw, corn cobs, dried corn stalks, cottonseed hulls, peanut hulls, sawdust, oat hulls, wood chips, wood shavings, pine shavings, used paper, sand, hemp, peat, and dry leaves. Different bedding has different pros and cons.

Straw is traditionally preferred bedding for livestock, such as sheep. Yet, hay is cheaper than straw since straw has many other uses. Sawdust isn't recommended for shelter that house wooled sheep since it can get into their fleeces. Peanut hulls and wood chips are also fine options as bedding, although they are less absorbent.

Shredded paper is both cheaper and more absorbent than straw. One negative is that it can make a huge mess if pieces of paper escape the shelter. Many sheep farmers prefer sand, depending on the availability and cost. You can check the availability and cost of different bedding materials and check how absorbent they are before choosing one. The trick is to pick an affordable bedding material that can keep the shelter dry and clean for the longest duration.

# Housing and Fencing Checklist

• Weigh up your perimeter and interior fencing options depending on the size of your land, flock, predation risks, sheep grazing methods, and budget.

• If your property has existing fences, evaluate their strength, and try to renovate them by repairing weak points and further strengthening them.

• Decide on an interior fencing mechanism that can be utilized when managing pasture and controlling sheep.

• Get to know about the climate and weather challenges in your area. Talk to local sheep farmers and learn about how they provide shelter to their sheep.

• Evaluate your land and see if its topography can provide enough shelter to your sheep about the climate and weather challenges they are bound to face in your area.

• Check if your property has any existing structures that can be used as housing for sheep and lambs. Renovate any such structures to be used to shelter sheep and lambs.

• Construct new housing for sheep if your management methods and local climate and weather require shelter for sheep.

• Come up with a housing strategy to quarantine and nurse sick sheep and any lambs born during adverse weather.

• Get to know the bedding options available in your area. Pick one or two bedding options depending on their availability and affordability to provide clean and comfortable housing for your sheep.

# Chapter 5: Sheep Nutrition and Feeding

Sheep, like other livestock, do not need specific food, but you must understand their essential nutritional needs, such as energy, protein, minerals, vitamins, fiber, and water. Different feedstuffs can fulfill these nutritional requirements. It's also important to make sure that sheep have access to balanced nutrition to ensure their health and wellbeing.

## Essential Nutritional Needs of Sheep

It's important for you to understand sheep and lambs' nutrient requirements clearly to ensure their health and wellbeing and the success of your sheep farming venture. These requirements vary with breed, age, sex, genetics, weight, level, and stage of production. If you're raising high-performing sheep, their nutritional needs are comparatively higher.

Sheep that aren't pregnant or nursing have lesser nutritional requirements. The level of nutrition that sheep receive generally dictates whether they gain body weight or lose it, so you can use their body weight to gauge how well their nutritional requirements are being met.

## Energy

Just like humans, sheep need energy to live and function. Most carbohydrates and fats in sheep's diet are spent towards energy production, while any excess proteins are also used for the same cause. Carbohydrates act as the major energy source and are usually found in grains, pasture and browse silage and hay.

Most sheep farmers struggle to make sure that their sheep's energy requirements are matched while avoiding over or underfeeding. Underfeeding leads to energy deficiency indicated by weight loss, reduced growth, and even death. Reduced conception rates, fewer multiple births, and less milk production are signs of energy deficiency in reproducing females.

Low energy consumption also leads to decreased fiber quality and wool growth. Sheep that do not receive enough energy through their diet are more susceptible to gastro-intestinal worms since the lack of energy weakens their immune systems.

The alarming outcomes of underfeeding do not mean you should overfeed your sheep. Excess energy consumption can lead to impaired reproductive function and pregnancy toxemia. Most consumers find overweight lambs undesirable, so overfeeding should also be avoided.

## Protein

The rumen of a sheep produces protein using amino acids. As a result, the quantity of protein in a sheep's diet is more important than its quality. The requirement of protein of sheep usually varies according to age, weight, and many other factors. Young and growing lambs and lactating ewes usually need more protein in their diets.

Most sheep farmers use soybean meal as a protein supplement. Sunflower meal, whole cottonseed, cottonseed meal, peanut meal, whole soybeans, fishmeal, rapeseed meal, and alfalfa pellets are less common protein sources. Legume hays harvested during their mid-bloom stage can also provide moderate levels of protein.

Farmers use protein blocks to ensure that their sheep receive enough protein. Although a convenient method, protein blocks are expensive. It's also difficult to control protein intake, and any surplus protein consumed by a particular sheep is used for energy, so there is some wastage and inefficiency associated with this method. Animals that regularly receive surplus protein also face negative health effects since excess is converted to ammonia and blood urea.

### Minerals

The diet of sheep should include sixteen essential minerals. They can be divided into two groups as macro-minerals and micro-minerals (also known as trace minerals). Macros-minerals are needed in high amounts, while micro-minerals are required in smaller amounts. Macro-minerals are sodium, calcium, chloride, magnesium, phosphorus, sulfur, and potassium. Micro-minerals are copper, iodine, iron, zinc, manganese, cobalt, molybdenum, fluoride, and selenium.

### Salt (Sodium and Chloride)

Sodium and chloride that salt contains play a vital regulatory function in sheep. As a result, it's important to provide sheep with enough salt. Salt deficiency can lead to reduced water and feed intake, growth, and milk production. Sheep licking dirt and chewing wood are common signs of salt deficiency. Salt is often used to regulate the feed intake of food and free-choice mineral mixes.

### Calcium and Phosphorus

These minerals play a vital role in developing and maintaining the skeletons of sheep. Deficiencies and imbalances of these minerals can lead to rickets and urinary calculi. Most fodders contain enough calcium. Sheep fed high-grain diets are at the risk of calcium deficiency, although such diets are rich in phosphorus. You're recommended to maintain a 2:1 ratio of calcium to phosphorus when feeding sheep.

### Vitamins

The diet of sheep should include essential vitamins such as vitamin A, D, and E. Although vitamin A is absent in plant food sources, beta-carotene synthesizes it. Consuming enough vitamin D ensures that sheep are safe from health conditions such as rickets and osteomalacia (softening of the bones). Likewise, vitamin K regulates blood clotting in sheep. B-vitamins are synthesized in the rumen, so they need not be included in the diet of sheep.

### Fiber

The healthy functioning of the rumen requires the bulk created by fiber. It increases salivation and rumination. Experts believe that sheep should have at least a pound of roughage in their diet. Chewing on wood or wool is a common sign of fiber deficiency in sheep.

### Water

Just like most animals, water is the most important nutrient for sheep, but many sheep farmers neglect this important aspect of feeding. Sheep generally consume between a half and four gallons of water every day. Environmental conditions and psychological factors play an important role in determining the daily requirement of water in sheep.

Sheep voluntarily drink water two to three times a day. The intake may increase in dry conditions or due to high-salt and high-protein diets. Water deficiency can lead to reduced growth and milk production. Sheep that consume enough water usually have lesser chances of developing digestive problems and urinary calculi.

# Nutrition During Breeding

The nutritional requirements of ewes usually remain the same during breeding. One exception is "flushing," a method practiced by sheep farmers to condition their bodies for breeding before or at the beginning of the breeding season. Feeding grain or moving them to a high-quality pasture helps accomplish flushing. Flushing has proven to increase lambing rates. Still, it's not required if the ewes are in good condition.

## Gestation

The nutritional needs of ewes during early and mid-gestation only slightly increase. The same feed can be continued during this stage, although you should make sure that nutrient deficiency is avoided. Poor nutrition during early and mid-gestation can affect embryo implantation and placenta growth.

The late gestation stage sees the ewe's nutritional requirements increasing substantially. Approximately 70% of the fetus's growth occurs during the final four to six weeks of pregnancy, so good nutrition is essential to ensure fetal growth and the production of milk, especially during the last two weeks.

During late gestation, nutrition deficiency can lead to various pregnancy diseases, increased postnatal losses, reduced birth weights of lambs, lesser mothering ability, and reduced milk production. In addition, the calcium intake of ewes needs to be increased during late gestation.

## Lactation

The nutritional requirements of ewes reach a peak during lactation. This is multiplied if the ewe gives birth to multiple offspring. Ewes that give birth to twins usually produce 20 to 40% more milk compared to those who nurse only one lamb.

### Ewe Lambs

The growing bodies of lambs and yearlings require more nutrition compared to physically mature sheep. Subsequently, you're recommended to make sure that lambs and yearlings' nutritional needs, especially ewe lambs, are met. It's also advisable to manage ewe lambs separately from mature ewes of your flock. You can let them join the flock once they have been bred for the second time.

### Feeding Lambs

There are different ways to feed lambs, depending on your plans for them. Most sheep farmers focused on gains provide concentrated feed to their lambs. Lambs that are pasture-reared also gain more and are less susceptible to worm problems if they are supplemented with some type of feed. Nutrient requirements of lamb usually vary according to their age and potential for growth. Frame size can determine the growth potential of lambs.

The temperature of the livestock's environment also plays a vital role in their growth. If the temperature is lower than sheep's critical temperature, they need to spend energy to maintain enough temperature in their bodies. The length of the fleece and the amount of fat in it usually determine a sheep's critical temperature.

For example, a recently shorn sheep's critical temperature is around 50° F. The critical temperature of a sheep with a 2.5-inch thick fleece is usually around 28° F. The nutritional requirement increases or decreases by 1% for every 1° F variant from the critical temperature. High-quality hay is recommended during colder months if the temperature is lower than the sheep's approximate critical temperature.

The activity level of sheep also affects their nutritional requirement. If your sheep are made to graze a large paddock and required to walk a long distance for grazing and drinking water, their nutritional requirements will be higher. Pan-fed sheep that receive little physical exercise, on the other hand, have lower nutritional requirements compared to free-ranging sheep.

### Feedstuffs

The most natural and economical diet for sheep is foraging, but you can also fulfill their nutritional requirements by feeding them a variety of feedstuffs. The rumen can also adjust to changes in diets and routines if it's given time to adjust. You can also adjust feedstuffs if you make sure that the nutritional needs are met, imbalances are avoided, and the health of the rumen isn't affected.

# Pasture, Range, Forbs, and Browse

These are usually the most economical ways of feeding sheep. They are also the most effective and convenient ways of ensuring that sheep's nutritional needs are met. Pasture contains high amounts of energy and protein. It's also highly palatable for sheep.

Though, high-production animals sometimes struggle to eat enough when vegetation when forage is wet or contains a lot of moisture. It causes loose feces that usually accumulate on their backs. Docking is usually practiced to overcome the issue. Pasture plants need to be managed well since their nutritional value, palatability, and digestibility decline as they mature, so it's recommended to rotate pasture or even clip is to keep pasture plants in the vegetative state.

Forbs usually offer more crude protein and better digestibility compared to grasses at certain stages of maturity. Most sheep often eat weeds over the grass. An increasing number of sheep farmers use hydroponic fodder such as barley sprouts to feed sheep for their nutrition value, although the moisture content is high.

### Hay

Forage that has been mowed and dried is known as hay. It's used as the primary food source for sheep during the winter and dry seasons when forage plants do not grow or thrive. The plant species and their level of maturity at harvest determine the quality of hay. Also, the quality of hay also depends on how well they are stored.

Hay offers moderate amounts of energy and protein for sheep. High-quality grass hay offers more nutrients than low or medium-quality legume hay. But high-quality legume hay offers three times as much calcium and 50 to 75% percent more protein than low and medium-quality grass hay.

It's important for you to realize that feeding the right type of hay is more important than feeding the best hay. It's safe to assume that hay that provides the most nutrients at the lowest cost is the "best" hay. The palatability of hay is as important as its nutritional value.

### Silage or Haylage

Silage is a livestock feed developed by fermenting high moisture herbage such as forage or grain crops. While silage can feed sheep, you need to practice caution not to feed your sheep moldy silage as it can cause listeriosis. Listeriosis sometimes causes miscarriage in ewes, so if you intend to feed your sheep silage, feed it before it develops mold.

### Baleage

Another type of feed that has become increasingly popular among sheep farmers is baleage. It is made using forage that has a high content of moisture in it. Then it's baled into round bales wrapped or sealed in plastic. Baleage is ensiled with 40 to 60% moisture; it is only 20% and over 65% in hay and silage.

### Concentrates

Feedstuffs that contain high amounts of nutrients are known as concentrates. If a forage diet lacks certain nutrients, you will be required to provide your sheep with concentrates to fulfill their nutritional needs. Feeding concentrates may be more economical than forages in small-scale sheep farming ventures. Concentrates can be divided into two types as carbonaceous and proteinaceous, or energy feeds and protein feeds.

### Energy Feeds (Carbonaceous)

These concentrates contain high amounts of digestible nutrients. But they are low in protein. Cereal grains such as barley, corn, oats, wheat, milo, and rye are common energy feeds. While cereal grains contain high amounts of energy, phosphorus, and other nutrients, they are low in calcium. Thus, you are advised to supplement calcium if you're feeding your sheep energy feeds. Also, higher concentrate diets need to be gradually introduced to sheep so it allows time for their rumens to adjust. Failing to do so can lead to many metabolic and digestive health problems.

### Protein Feeds (Proteinaceous)

These feedstuffs contain high levels of protein. Cottonseed meal, soybean meal, and fishmeal are common protein feeds. Soybean meal is the most economical and common protein feed found in the United States. You're advised to watch the intake of protein feeds since sheep do not increase productivity with excess protein or store it for later use.

### Commercial Feeds

Various companies offer sheep feed balanced for the specific needs of sheep. They usually come as textured or processed and focus on specific age groups and product categories. You can feed your sheep with commercial feed. But it should not be mixed with other grains as it can cause imbalances in nutrition.

### Pelleted Supplements

These supplements result from combining different feed ingredients to control feed costs such as soybean meal, corn, and minerals. Commercial pelleted supplements also contain high levels of protein, and various essential minerals and vitamins. They can be easily combined with whole-grain diets.

## By-Product Feeds

Many by-products are fed to sheep to cut down feeding costs and wastage. These by-products result from various processes that develop various agricultural products. A good example is corn gluten meal, which is produced through the corn milling process. Similarly, the production of soy oil and soy meal creates soybean hulls. Wheat middling, beet pulp, brewer's grain, and citrus pulp are other examples of by-products. The nutritional value of by-products can vary. Hence, it's advised that you check nutrient contents if you're planning to feed any by-product feeds to your sheep.

## Mineral Sources

Sheep nutritionists believe that sheep can't determine the ideal minerals they should consume except for salt. Respectively, if you free-feed certain minerals, the sheep may consume more or less of those minerals rather than consuming the required amounts.

Various methods can be adopted to ensure that sheep receive enough nutrients, especially minerals such as salt, calcium, and phosphorus. If you're feeding rations, you can easily feed the ideal amounts of minerals to sheep, but rationing is only practical with smaller flocks due to the difficulty in making sure that all the sheep receive the required amounts of minerals. If you're the owner of a larger flock, you can mix essential minerals with loose salt to make sure that every sheep consumes the required amount.

# Sheep Nutrition and Feeding Checklist

• Consider your flock size and decide what type of feeding methods you will use.

• Evaluate the size of the land available to you and the food that it can supply.

• Clearly understand your sheep's nutritional needs based on their breed, age, production level and stage, physical activity, and climate they are in.

• Evaluate the feed available in your property and your area. Look into their nutritional properties and create a list of feedstuffs that you can use.

• Decide on the primary feedstuffs based on availability and affordability. Make sure that the feedstuffs that you choose provide the required nutrition to your sheep.

• Create a plan to provide enough nutrition for your sheep during the winter and dry seasons according to your area's climate.

• Choose a practical method to provide essential minerals such as sodium, calcium, and phosphorus for your sheep.

• Seek advice from local sheep farmers and check what is being fed to sheep, especially in ventures like yours.

# Chapter 6: Guarding Your Sheep

Many farmers in the United States lose sheep and lamb due to predation. How much of risk predators pose for your sheep usually depends on your farm and your region. Maybe you may not face that much of an issue with predators if you raise backyard sheep and live in an urban or suburban area.

Small farms in rural areas are more likely to run into trouble with predators. Though, the threat posed by predators varies from one region to another and depends on the common predators in those regions. Even if your farm is in an area with a lesser threat of predation, it's highly recommended that you don't take chances since even rare occurrences can cause a lot of damage to your flock.

There are many ways to control predators and minimize threats posed by them. Non-lethal methods of predator control are the easiest, although they may require a sizable initial investment. But lethal methods may not be attractive to you, although they are usually effective. Besides, using animals for guarding sheep usually requires some work with varying levels of success.

### Non-Lethal Predator Control

Environmentalists, animal welfare advocates, and even many farmers favor these methods since they do not involve harming the predators. Additionally, these methods do not put other animals, including sheep, in harm's way. Non-lethal predator control is practiced by almost every sheep farmer and can be highly effective depending on the type of methods employed and their quality.

### Fencing

Predators can be controlled with good fencing in most cases. It acts as the first line of defense against all types of harmful elements beyond it, including predators. Predators, however, will try various tactics to get to your sheep. These include jumping over or between the wires, digging under the fence, crawling through gaps in the mesh, and using physical strength to damage the fencing.

Sheep farmers prefer woven wire fences in regions where there is a high risk of predation despite their cost. If you live in such an area, you're advised to use woven wire fencing with stays only six inches or less apart and horizontal wires just two to four inches apart.

Another effective predator-proof fencing option is high-tensile fences. If your region has a high risk of predation, you're advised to use at least five high-tensile wires when fencing. The more strands you have on your fence, the more effective and expensive it will be. The bottom wires need to be placed closer together than the wires at the top. It's advised that high-tensile fencing has a good combination of both live and ground wires.

### Livestock Guardians

Animals such as dogs, donkeys, and llamas are popular among sheep farmers for guarding flocks against predators. According to the United States Department of Agriculture's National Animal Health Monitoring System (NAHMS), approximately 45% of sheep farmers in the United States employ livestock guardian. Of those guardian animals, approximately 30% are dogs, 14% are llamas, and 11% are donkeys.

## Guardian Dogs

The most popular livestock guardian animals are dogs. They have been used to protect livestock such as sheep from predators for centuries, with dog breeds specifically bred to guard sheep originating from Europe and Asia. The Great Pyrenees, Akbash and Anatolian Shepherd, Komondor, Polish Tatra, Maremma, and Mastiff are good examples of dog breeds bred to protect livestock.

Guardian dogs usually have large bodies either fawn or white-colored. They also have dark muzzles. Experts believe there aren't that many differences in terms of effectiveness among breeds. Livestock guardian dogs stay near sheep to repel any predators that may try to harm them. They usually work best as pairs, although one guardian dog can protect a small flock of sheep grazing a smaller pasture.

Factors such as sex and neutering or spaying do not seem to affect dogs' guarding capabilities. In contrast, genetics and training play the most vital roles in determining how effective a guardian dog is. They should be raised alongside sheep from puppyhood while minimizing human contact to make them bond with sheep more. Due to this, it's common for most guardian dogs to consider strangers as threats to livestock.

## Llamas

Not all sheep farmers, especially beginners, may be able to afford a good guardian dog or train one. However, a single llama can guard sheep without requiring any training. Llamas need to be kept single since having more than one would encourage them to bond with their own kind instead of the sheep.

Llamas are highly effective against dogs and coyotes. Female llamas or castrated males are more preferred over intact males since they might try to mate with the ewes at times, and that can cause injuries and even death. Llamas can be introduced to a flock of sheep in a smaller pasture, and they need not be raised with sheep from a young age, unlike guardian dogs.

Another advantage of using llamas to protect sheep from predators is that their diet is very similar to sheep. They do not need special feeds or shelter. They can be left alone with the sheep to look after themselves and the sheep with no special care or attention. Llamas also live long, usually between 15 and 20 years.

### Donkeys

Another great option for guarding sheep is donkeys. They have great herding instincts and naturally dislike dogs and coyotes. Donkeys easily bond with sheep and do a great job for protecting them from predators. A donkey can be made to bond with sheep by housing it next to sheep for around two weeks if it hasn't been raised among sheep.

The best combination for guardian donkeys is a duo of jenny and foal. A jenny alone can also do a considerably good job. Geldings may not be as aggressive towards predators as jennies and foals. Despite that, many sheep farmers prefer them for their desirable temperaments. Intact male donkeys aren't usually used as guardian donkeys due to their aggression towards both sheep and their owners.

Every donkey doesn't make a great guardian donkey. Some can be too aggressive towards sheep. Donkeys are best recommended for smaller flocks with less than 100 ewes. Donkeys are also advantageous because their upkeep is low.

### Cattle

Some sheep farmers graze sheep alongside cattle as a *flerd* (flock and herd) to protect them from predators, but getting the two types of livestock to bond well together can sometimes be difficult. When bonded, sheep will seek protection from the cattle whenever they are threatened. However, when sheep have not bonded enough with the cattle, they will stay away from the cattle, and they will be left largely unguarded.

# Proper Management Against Predators

Most animals that prey on sheep operate at night. Thus, keeping sheep near buildings can sometimes deter predators, especially if they are lit. If you raise backyard sheep, penning them in a well-lit building or yard can reduce the predation threat.

Scavenging is the stepping-stone to predation for some animals, such as coyotes. Because of this, sheep farmers are advised to properly dispose of dead livestock. Doing so can avoid encouraging potential predators to prey on sheep.

Lambs are more prone to predators than mature sheep. Therefore, shed lambing is advised instead of pasture lambing if possible. Areas of a property with a history of predation should be avoided as pasture during lambing. Likewise, flatter terrain should be preferred since predators won't use terrain to their advantage.

### Frightening Devices

Shepherds, for centuries, have used frightening devices to scare away predators. Such devices have evolved over the years from simple scarecrows to modern electronic guarding systems. Modern electronic guards use both sound and light as scare tactics. Multiple guards are required for large pastures, and sensor-powered ones are preferred since they automatically turn themselves on at nightfall.

### Plastic Collars

Most predators attack the throats of their prey. Plastic collars are used to protect those areas, especially of lambs, to prevent predators from causing life-threatening injuries to them. Plastic collars are easy to fit and remove and are usually kept on lambs until they are at least a year old. If you consider plastic collars, you will need to ensure that they are adjusted every three months.

## Lethal Predator Control

These predator control methods usually result in either injuring or mostly killing the predators. They should be employed only if non-lethal control methods prove unsuccessful. Federal and state laws do not allow killing certain predators such as Golden Eagles, Bald Eagles, and bears, although it's usually legal to kill foxes, coyotes, and mountain lions.

### Shooting

Hunting is often practiced to control the populations of certain predators, such as coyotes. A lesser population usually results in lessening the threat of predation. A lesser population also means there will be enough natural food supply for such species that also decreases predation. Conversely, this method is most effective when carried out on a wider scale instead of you as a small-scale sheep farmer shooting a few coyotes that won't do much to reduce their population in your area.

### Trapping

One of the most effective ways of controlling certain predators is through trapping. Common trapping methods include snare traps and leg holds. The downside of trapping is that it can injure not only target species but others, even protected ones. Consequently, certain trapping methods such as leg-hold traps are banned in some states and are also banned in 88 countries throughout the world.

### Livestock Protection Collar

These collars provide a lethal poison to predators such as coyotes that attack lambs wearing the livestock protection collars. The poison causes death within two to seven hours. The lethal solution also contains a yellow or pink dye that helps identify animals that are contaminated.

Livestock protection collars have proven to succeed in areas with high predation, and were more humane methods have failed. Sheep farmers in ten states in the United States may use the method in cooperation with the USDA Wildlife Services. These states are Texas, South Dakota, New Mexico, Montana, Virginia, Utah, West Virginia, Ohio, Pennsylvania, and Wyoming.

### M-44 Cyanide Injector

This contraption consists of an ejector that releases cyanide power into the predator's mouth when it pulls back on the baited unit. The cyanide powder dissolves with the moisture in the predator's mouth that creates hydrogen cyanide gas that kills it within 10 seconds to two minutes. The M-44 should be used in cooperation with the USDA Wildlife Services.

# Predator Control Checklist

• Discover the types of predators in your area by talking to local sheep farmers.

• See if your fencing is strong enough to protect your sheep from predation.

• Identify any weaknesses and weak points in your fencing and consider installing frightening predator devices.

• Try penning your sheep in an area where there is artificial light at night. However, this option may not be practical for larger flocks.

• If fencing and predator frightening devices fail, consider using guardian animals such as dogs, llamas, or donkeys depending on your budget and expertise in training.

• If your area has a high predation history, fit plastic collars on lambs until they become yearlings.

• If the above non-lethal predator control methods fail, you may need to consider lethal predator control methods such as shooting, trapping, livestock protection collars, or M-44 cyanide injectors.

• Discover any protected predatory animals and any predators that your state laws prohibit you from injuring or killing.

• You're advised to consult with experienced local sheep farmers and the USDA Wildlife Services before employing lethal predator control methods.

# Chapter 7: Sheep Shearing, Care, and Maintenance

Your sheep require care and maintenance to remain healthy and productive. Shearing involves the removal of wool. Breeds of sheep that produce wool need to be sheared at least once a year and crutched before the lambing season. Sheep also require maintenance, where you will need to pay attention to their hooves. Plus, identification and record-keeping also make caring for and maintaining sheep more efficient and successful.

## Sheep Shearing

If you're raising breeds of sheep that grow wool, they need to be sheared at least once every year. Shearing usually takes place in the spring. It's timed that way since the coats provide warmth during the winter and are removed before they overheat the sheep during the summer. Not to mention, it's preferable to have sheep sheared before lambing.

Shearing makes it easier for lambs to drink from their mothers. Sheared sheep also take up much less room in shelters that might be used for lambing. If you shear your sheep during the winter or early spring by any chance, it's important to provide them with extra nourishment since they will need to spend more energy to maintain body heat.

### Shearing

You may think that shearing is a job that anyone can do. While you may be correct, everyone isn't a good shearer. It's considered a sought-after skill. The sheep need to be handled and shorn without injuring them. Improper shearing cannot only harm your sheep but also decrease the quality of wool sheared.

Most small-scale sheep farmers find it difficult to locate professional shearers. Taking sheep to a shearing shed might be a better option if you arrange before transporting your sheep. Large-scale sheep farmers enjoy the services of shearing crews who usually come along with a trailer that can accommodate shearing.

Although shearing requires skill, it's a skill that can be mastered with some hard work. You can attend a shearing school to learn more about the art of shearing. Upon completion, you can shear your own sheep and attend shearing competitions.

### Shearing Equipment

Good shearing requires good shearing equipment. They make shearing much safer and easier for both the shearer and the sheep. Most shearers use electric cutters. They come at many prices. A good electric cutter can cost you somewhere between $250 and $500. It's also important to keep a few extra cutters with you since they can become dull. Shearing with dull tools is highly advised against since it can be dangerous.

### Preparing Sheep for Shearing

First, arrange to have your sheep sheared. If you're planning to have a professional shearer do the job, make an appointment beforehand. If you are taking your sheep to a shearing shed, you will need to gather them up and take them, being sure to confirm with the shearer.

If you are doing the shearing yourself or if a shearer is coming to your property to do the shearing, have your sheep penned. Divide rams, ewes, lambs, and yearlings into separate groups according to their breed or grade. It's also recommended that you refrain from feeding your sheep before shearing so that the shearing floor is much cleaner.

Sheep sometimes find shearing discomforting if their stomachs are full. Sheep should be dry when shearing, and it's important to keep the shearing floor clean and dry. It should also be swept after shearing each sheep so that a clean surface is available for the next one.

### Skirting Fleeces

Tags and belly wool need to be separated from the rest of the fleece after shearing. Lay the fleece with the flesh side facing down. Remove any wool that is off-color or dirty, short or matted wool, tags, and any other contaminated areas. Finally, roll the two sides of the skirted fleece toward the middle and nicely roll it from one end to the other. The flesh side should be facing out. Skirting is an important skill to have as it makes wool more appealing to hand spinners and other buyers.

### Packaging Wool

You can package fleeces in plastic garbage bags or cardboard boxes. It is important not to use poly feed sacks and burlap bags to package wool since they can contaminate it. Large square bales are used when packing a lot of wool. Clear plastic wool bags are the preferred packing option for wool.

Package tags, belly wool, off-color, seedy, burry, cotted, chaffy, stained, and dead wool separately. Correspondingly, any black wool should be kept and package separately from white wool. All bags containing wool needs to be labeled according to what they contain. Store wool in a dry and clean place before they are sold. Properly sorting and packaging wool can cause your wool selling for a higher price.

### Hair Sheep

If you have hair sheep in your backyard or small farm, they should be kept separately from wool sheep. The existence of hair fibers in wool fibers reduces wool quality. Any hair wool crosses need to be the last sheep to be sheared. Their wool should be kept separately from the fleeces of wool sheep.

### Crotching or Crutching

Crutching is a type of shearing where only the area around the vulva and udder are sheared. If you are shearing ewes before lambing, you can get them crutched simultaneously. Crutching makes sure that the area is dry and doesn't attract disease like a blowfly strike.

# Sheep Hoof Care

One of the most important parts of sheep management is hoof care. Various hoof diseases can have significant negative effects on the health and productivity of sheep. Therefore, you're advised to regularly check your sheep's hooves for signs of diseases and excess growth. Culling might be required if your flock includes sheep that ail from recurring hoof diseases and excessive hoof growth that do not respond to treatment.

### Hoof Trimming

Factors such as genetics, breed, nutrition, management, and soil attributes such as moisture content determine hoof growth. Sheep that graze on soil with high moisture content and free of rocks need regular hoof inspections. Also, sheep housed may need more hoof maintenance than free-range sheep.

When trimming hooves of sheep, first remove any dirt, stones, or manure stuck in them. If you notice a rotten smell, it might be a sign of foot rot. The dirt and junk can be cleaned out using a small knife. The perimeter of the hooves then needs to be trimmed.

Trimming should be stopped at the first sign of pinkness. Trimming should start from the heel and go towards its "horny" area. You can familiarize yourself with what a properly trimmed hoof should look like by inspecting the hooves of newborn lambs. Hoof trimming can be combined with other management tasks such as vaccinating or shearing and should not be carried out in hot weather or late gestation.

# Diseases Affecting the Hoof

Many diseases can affect the hooves of sheep. Lameness is a sign of hoof diseases and should not be ignored since some can be very serious.

### Bluetongue

Biting insects usually contribute to the spread of this viral disease. It's a non-contagious disease that can be easily diagnosed by identifying the presence of a reddish or brown colored band around the coronet.

### Foot Abscess

Swelling of the soft tissues located immediately above the hoof and draining abscesses in the areas between the toes are signs of a foot abscess. It's caused by the bacterial infection of food tissue that is damaged. Foot abscess mostly affects the front legs and can be treated with anti-bacterial compounds.

### Foot Rot

This is one of the worst diseases known to the sheep industry around the world. It causes loss of production and the need to cull animals. It also costs a lot of money to provide treatment, especially for labor. Foot rot, a disease caused by two anaerobic bacteria, is present wherever there are sheep.

The bacteria that cause foot rot are introduced to farms through infected animals. Once introduced, they spread in warm and moist conditions. Though, these bacteria can only last in the soil for two to three weeks.

Healthy sheep that walk on the ground or manure infected with the bacteria can develop foot rot. It will then continue to infect the entire flock unless the infected animals are culled, or the bacteria are effectively destroyed using early diagnosis and treatment.

Maintaining properly trimmed hooves is one of the best ways to avoid the development of foot rot. Still, excessive or aggressive trimming should be avoided as it can aid the development of the disease. It's also advisable to soak the feet of sheep in a 10% zinc sulfate solution.

Vaccination can also help prevent the development of foot rot, but it does not guarantee complete immunity from the disease since vaccines do not cover all the foot rot strains. Any sheep that do not respond to treatment or have a history with the disease need to be culled. However, it is more economical to control the disease through prevention since vaccination can be expensive.

Always treat each new addition to your flock as a sheep infected with foot rot. Isolating new animals for a month, trimming their hooves upon arrival, treating the feet upon trimming, and regular inspections of the feet during quarantine are great ways to safeguard your healthy flock from introducing foot rot. You're also advised not to purchase sheep that have been infected with foot rot and not to buy animals from sale barns.

### Foot Scald

White, blanched, red, and swollen tissue between the toes of sheep is a sign of foot scald, which is a non-contagious disease. It's much easier to treat compared to foot rot. Placing sheep on drier pasture, the topical use of copper sulfate or zinc sulfate, and footbaths of 10% zinc sulfate solution can easily treat foot scald.

# Sheep Identification and Record-Keeping

It does not matter how small your flock of sheep may be; proper identification and record-keeping make things much easier and contribute to your venture's success. Record-keeping is a great way to identify the ideal lambs to keep as replacements, best-performing ewes that need to be kept and those who need to be culled, and which rams produce the best lambs.

Besides, the National Scrapie Eradication Programs require you to maintain records regarding animal disposition. Maintaining records regarding health products such as antibiotics and anthelmintics is also becoming increasingly important.

### Animal Identification

You must first identify sheep in your flock so record-keeping can begin. You can choose from different identification methods depending on the number of sheep in your flock, your preferences, needs, and budget. Sheep identification should ideally be loss and tear-resistant, easy to apply, and easy to read and understand.

### Ear Tags

Many sheep farmers use ear tags to identify their sheep. There are all sorts of ear tags made in various sizes, designs, and materials such as aluminum, brass, and plastic. Rotary tags, button tags, looping tags, and swivel tags are the most common in the United States.

Brass tags are perfect for newborn lambs as they are lightweight. One minor difficulty is that you will need to catch lambs to read their tags. Although metal tags are a cheaper option, they can easily be ripped out of the ears, which can cause infections. Looping and swivel tags are not only durable but also more readable than brass tags. Temple tags usually have the least risk of ripping out, but you need to punch a hole in the ear to insert the tag.

### Scrapie Identification

The United States Department of Agriculture (USDA) has made it mandatory for all sheep and lambs to have proper identification in the form of ear tags before leaving their farms, despite the flock size. Every ear tag should carry the premise identification number and the sequential identification number. You can use the sequential number for your record-keeping purposes. And the USDA dictates you keep records of sheep for five years after they are sold.

You can get your premise identification number and metal ear tags for free by calling the toll-free number 1-866-873-2824. That said, the USDA does not provide free plastic ear tags unless you're requesting them for the first time. Sheep farmers participating in the Scrapie Flock certification program need to identify all their sheep that are older than a year using a tamper-proof identification method such as ear tags, microchips, or tattoos.

### Tattoos

Although difficult to reach without catching the sheep, tattoos are a permanent identification method. Tattooing an animal does not decrease its value. Once they are tattooed, the identification remains for their lifetime.

### Ear Notching

This method is more of a differentiation method rather than identification. For example, some sheep farmers use ear notching to easily identify sheep that need to be culled.

### Electronic ID

Many sheep farmers are starting to use electronic IDs to easily identify their sheep. The method uses radio frequency identification technology. The ear tag used in this method contains a microchip and a tiny copper antenna within the tag. Electronic ear tags can be easily applied and removed.

# Sheep Shearing, Care, and Maintenance Checklist

- Evaluate how often your sheep need to be sheared depending on their breed and purpose.

- Decide how you will have your sheep sheared. If your flock is small, you might need to visit a shearing shed. If you have a larger flock, you might need to ask shearers to come to your small farm.

- Get in touch with local sheep farmers and shearers to find the best way to have your sheep sheared.

- If you're interested in learning to shear, find a good shearing school you can attend.

- Create a schedule to carry out hoof inspections and trimming.

- Check hoof rot and other similar diseases are common in your region.

- Come up with a plan to quarantine and treat any sheep that have signs of hoof diseases and ways to stop the disease from spreading.

- Create facilities to quarantine any new sheep you purchase for a month before inspecting them and introducing them to your flock.

- Consider different sheep identification methods and choose a suitable method depending on your budget and needs.

- Maintain records of all the sheep you own. It's highly advised that you maintain records electronically while maintaining back-ups of your original files.

# Chapter 8: Sheep Reproduction: Understanding Rams and Ewes

The next level of raising sheep in a backyard or small farm is breeding them, but you may not need to breed sheep if you are raising them for wool. However, if you're raising sheep for meat or milk, you need to breed your sheep to sustain your flock. Breeding takes a lot of work. Breeding requires you to accommodate an intact male associated with certain risks.

There's also the option of seeking the services of a ram owned by someone else. Either way, breeding is a viable option for anyone who raises sheep and wants to expand their flock without purchasing new animals.

Breeding should be carefully done upon careful consideration of your purpose. Are you looking to maintain optimal breed standards? Or are you trying to produce lambs with desirable characteristics belonging to different breeds? Breeding is a complex topic, and first make yourself familiar with it before you proceed.

# Reproduction in the Ewe

Ewes are considered to have reached sexual maturity upon having their first heat. Genetics, breed, nutrition, size, and birth season usually determine the age of puberty. Ewe lambs usually reach puberty when they are between five to 12 months old. Ewe lambs born in the spring usually reach puberty in their first fall, and lambs that were born later may take longer to do so.

And single-lambs reach puberty much sooner than twin or triplet-born ewe lambs. Meat and hair sheep breeds also reach puberty much sooner than wool breeds. It's also normal for crossbred ewe lambs to reach sexual maturity sooner than purebred ewe lambs.

## The Estrus Cycle (Heat Cycle)

The estrus cycle's length that regulates the reproduction in sheep usually varies from 13 to 19 days. It consists of four phases, known as proestrus, estrus, metestrus, and diestrus. Ewes usually respond to rams and mate during the estrus. It usually lasts for around 24 to 36 hours.

The ovulation usually occurs during mid or late-estrus. It's when the ovary releases eggs. Metestrus usually lasts for three days and begins as the estrus ends. The corpus luteum that maintains pregnancy in ewes' forms during metestrus. The corpus luteum is fully functional by the time diestrus begins.

Proestrus usually starts as corpus luteum regresses and remains until the start of the next estrus. It usually lasts for nine to eleven days, where rapid follicular growth occurs. The state where the normal cycle stops is known as the anestrus.

Different seasons affect estrous cycles depending on the number of hours that the sheep's eyes are exposed to daylight. Most sheep reach estrus as the length of the day decreases. As a result, October and November are the most natural times to breed sheep in the United States and Canada.

Breeds such as Dorset, Marino, Rambouillet, Karakul, Finnsheep, and hair sheep breeds are less seasonal and can breed throughout the year or have extended breeding seasons. Flocks located closer to the equator usually have longer breeding seasons compared to others.

It's usually difficult to read the signs of estrus in sheep unless a ram is around. Mature ewes in heat usually seek the ram and stand still, allowing him to mount them. They may also wag their tails fast or even try to mount the ram. However, young ewes rarely exhibit such behaviors.

### Gestation

This stage usually lasts for around 150 days upon mating. It's important that you keep track of the dates so that you can start feeding ewes with a grain a few weeks before they are expected to start lambing and carefully observe them during the last days of gestation. You're advised to start with one pound of grain per ewe and gradually increase the amount. The maximum amount of grain that should be fed during late gestation usually depends on the breed.

The ewes also need to be crutched as the gestation period ends. Vaccines also need to be administered four weeks before the expected date of lambing. You can also deworm your ewes around the same time.

# Reproduction in the Ram

You're advised to treat the ram as the most important member of your flock. Sadly, many sheep farmers neglect the ram, although he's the one who contributes most of the genetics to their flocks. The wellbeing of your ram will contribute to breeding and the eventual success of your sheep-raising venture.

### Puberty

The age at which a ram's reproductive organs become functional, he's ready to mate, and secondary sexual characteristics develop, known as puberty. Ram lambs usually reach puberty when they are five to seven months old and upon achieving 50% to 60% of their mature body weight.

The time it takes a ram to reach puberty usually depends on genetics, breed, and nutrition. Prolific, meat, and hair breeds usually reach puberty much sooner than other breeds of lambs, wool breeds, and especially those who consume low-nutrition diets.

### Spermatogenesis

The production of the male reproductive cell or sperm is known as spermatogenesis. It usually takes about seven weeks for rams to produce sperm. Experts believe that the larger scrotum size and firm tail are signs in rams that usually indicate good sperm production and reserves.

Nutrition plays a vital role in sperm production. With that in mind, experts recommend that you provide a nutrient-rich diet to rams, especially two months before breeding takes place. Although, avoid overfeeding since it can reduce sperm production in rams.

# Seasonal Effects on Reproduction

Sheep that live in more temperate climates are more seasonal breeders, although they are less affected compared to ewes. A ram's ability to mate and produce sperm usually varies depending on the season of the year. Their breeding capabilities reach a peak in the fall, where the traditional breeding season takes place.

Some breeds such as Rambouillet, Dorset, Polypay, Merino, Romanov, Finnsheep, and hair breeds are usually less seasonal to breed capabilities and behavior. Temperature plays a vital role in determining the fertility of rams. For example, experts believe that even a half a degree variant in body heat can affect libido and spermatogenesis.

## Mating

Ewes in heat usually seek the rams. They sniff, chase, and even try to mount the ram. It's normal for rams to fail a few times before mounting the ewe, and they may also mate with the same ewe more than once. Some rams select older ewes or ones of their own breed over younger ewes and ones of other breeds.

Some sheep farmers only use one ram for a group of ewes. When multiple rams are used, the older rams may dominate the younger ones and hinder less dominant rams from breeding. Such behavior can cause fights. Another issue is that it is difficult to identify which rams are superior or inferior at breeding in multi-sire scenarios. Therefore, a single ram is recommended for smaller groups of ewes.

## Libido

A ram's willingness to mate is known as libido. The ram's testosterone level usually determines its libido. Different rams have different libidos. Some can be seasonal while others maintain the same libido throughout the year. Age and health are also determining factors of libido.

Some rams naturally inherit poor libidos. According to experts, some rams are homosexual and may refuse to mate. A ram's libido can be determined using a Serving Capacity Test. It involves exposing the ram to estrus ewes and recording their mating activities over two or more weeks. Serving capacity tests are often used to identify high-performance rams.

You can also simply determine a ram's libido by exposing him to estrus ewes and carefully observing its mating behavior. A marking harness or raddle paint can be used for careful monitoring. The marking crayon of the harness or the paint's color should be changed after 17 days. A ram that fails to mate with ewes, even after marking and monitoring, needs to be replaced with a better-performing ram. If a ram re-marks ewes after a 17-day cycle, it's an indication of a sterile or sub-fertile lamb.

# Ram Management

It's natural for a lamb to lose approximately 15% of his body weight while breeding. That's why you're advised to ensure that the rams are in optimal physical condition prior to breeding. Thin rams struggle to impregnate ewes, while fat rams may be too lazy to breed or sub-fertile, especially in hot weather. Rams should be sheared, dewormed, put on a high-nutrient diet, and have their hooves trimmed two to four weeks before breeding begins.

### Ram to Ewe Ratio

Breeding experience, age, pasture-size, terrain, and the number of ewes in the group are factors that determine how many ewes a ram can breed during a breeding season either 34 or 51 days long.

A good ram can usually mate with three to four ewes every day. Experts recommend that you use one ram per 35 or 50 ewes. Using one ram per 100 or even 150 sheep is also common. The percentage of rams becomes higher for larger flocks.

# Breeding Systems

The systematic approach where the genetic value of livestock is evaluated and then taken forward is animal breeding. A breed of sheep is a group of sheep showcasing homogeneous characteristics, appearance, and behavior distinguishable from other sheep.

### Pure-Breeding

Facilitating the mating between sheep of the same type or breed is known as pure-breeding. Such a flock can be easily managed as a single flock since all rams and ewes belong to the same breed. Pure-breeding aims to safeguard superior genetics considered valuable.

Improvements to genetic traits of purebred sheep need to be documented with performance records. The National Sheep Improvement Program (NSIP) collects that data. In return, breeders receive across-flock Expected Breeding Values or EBVs.

An EBV provides estimates the genetic merit of a particular animal in relation to a particular genetic trait. It provides a comparison between the expected performance of the particular animal and the average performance of that trait within the breed.

## Out-Breeding

When animals of the same breed, which are at least four to six generations apart, are bred, it's known as out-breeding. It is the more recommended breeding practice.

## Inbreeding

When closely related animals are bred, it's known as inbreeding that includes son to the dam, sire to daughter, and brother to sister breeding systems. This breeding is focused on the higher frequency of the pairing of similar genes. Inbreeding is only recommended for qualified operators. You should avoid allowing or facilitating inbreeding within your flocks.

## Linebreeding

This breeding involves sheep that aren't closely related as they are in inbreeding. Instead, it may look to breeding systems consisting of mating between cousins or half-siblings to ensure that their offspring are related to a highly priced ancestor.

## Crossbreeding

When rams and ewes from different breeds or types are mated, it's known as crossbreeding. It's a systematic approach where desirable genetic resources are made to create commercially valuable and productive offspring. Breed complementarity and heterosis are the main advantages that are crossbreeding offers.

## Breed Complementarity

Different breeds have distinct strengths and weaknesses. Breeding systems usually aim to maximize the strengths and minimize weaknesses using superior genetic traits. For example, if you want offspring with great reproductive efficiency, lower maintenance, heavy-muscled, and fast growing, it will be hard to find a breed that matches all those traits. However, you can crossbreed Polypay ewes are known for their moderate maintenance and reproductive efficiency with Suffolk rams, which are more heavy-muscled and fast growing.

### Heterosis

The term heterosis refers to the superiority of offspring compared to their parents that were crossbred. It's measured by the difference between the offspring's performance and the performance of their purebred parents. Crossbreds are more fertile, fast growing, and vigorous compared to their purebred parents.

# Sheep Reproduction Checklist

• Consider your purpose for breeding sheep. Explore if you have space, time, and resources to successfully breed sheep.

• Evaluate the profitability of breeding sheep and how you will market the offspring.

• Decide on the breeding system you are going to use to breed your sheep. Different types of breeding systems are suitable for different purposes of raising sheep.

• Decide whether you have space, time, and resources to raise a ram. The ram should be looked after well and kept healthy for breeding. Rams are also more difficult to handle compared to ewes and wethers.

• If you don't have enough land to raise a ram, you can seek the services of a good ram. Talk to local sheep farmers regarding rams available for crossing.

• If you are bringing a ram into your farm for breeding, make sure that he arrives with ample time to quarantine him to ensure that he's not carrying any diseases.

• A ram requires a dedicated pasture until ewes are ready to mate. Have a pasture that is large enough and has enough food for the ram.

• Make sure that the ram's diet is nutrient-rich starting from two months before breeding while making sure that they are not too thin or fat by the time breeding begins.

• Trim the hooves, deworm, and shear the ram two to four weeks before breeding.

• Observe the behavior of ewes and identify when they have reached estrus. Let the ram into the pasture that the ewes are in.

• Monitor how well the ram mates with the ewes. If you own the ram, it may be beneficial to keep track of his performance by using cradle paint or a marking harness.

• A single ram can mate with as many as 150 ewes. If you have a larger flock, you may be required more than one ram.

• Once mating is completed, and the ewes reach metestrus, remove the rams.

• Gestation usually lasts for 150 days upon mating. Keep track of the dates that the sheep mated so you can estimate when lambing will begin.

• Start grain-feeding ewes a few weeks before the expected date of lambing. You're also advised to crutch and deworm ewes before late gestation.

# Chapter 9: The Lambing Process

Thanks to evolution over thousands of years, most ewes have little difficult lambing. The process is regulated by a sequence of hormonal changes guiding the lamb to decide when time to be born. You must have a thorough understanding of the lambing process so you know when ewes are about to give birth and provide help if necessary.

As a ewe gets closer to lambing, she may stop eating, her teats and udder will be swollen, and vulva dilated. First-time mothers will find the process a little confusing, especially yearlings. You need to be prepared when ewes get close to lambing. You should have lambing facilities ready along with lambing supplies. Ewes also need to be carefully managed and fed, leading to lambing.

## Pasture vs. Shed Lambing

Both these methods have their own pros and cons. It's up to you to weigh those pros and cons in relation to your operations so that you will be able to provide the ideal conditions efficiently and economically to safely deliver lambs. Feed costs, climate, labor availability, predation, risk of disease, and market highs and lows are key factors that you will need to consider.

# Shed Lambing

Sheep farmers in the United States who own small flocks usually practice shedding lambing. So, it's suggested that you should follow the same approach after evaluating the pros and cons associated with it. Shed lambing allows early or out-of-season lambing that can usually fetch higher revenues. Lambing will take place in a structure such as a barn where even winter lambing can be carried out.

You can manage lambs more when delivered in a shed compared to pasture. You can control losses more effectively. Tasks such as treating ewes and lambs, vaccinating, and weaning become much simpler.

A lambing barn should generally have the capacity for 10% of your flock. Some ewes can be housed in the barn until the grass is available outside, while others can be moved outside upon successful lambing. Most lambing barns are constructed with drop pens. The availability of a few drop pens lowers the chances of ewes stealing lambs or mismothering.

Most lambing barns also feature jugs. Each jug can house a ewe and her lambs. Tasks such as deworming, tagging, and banding can take place in each jug while each family is confined. The families are then sent into a mixing pen after 24 to 48 hours, where they will bond with the flock.

Sometimes a few lambs will need bottle-feeding so it's important that a lambing barn or shed has a dedicated area for it. Shed lambing makes things easier for you since you'll be able to easily take lambs from their mothers if they can feed no lambs. You will have the luxury of either giving the lambs to another ewe or keep the family in a jug longer to make certain that they are doing well.

The downside of shed lambing is the high initial investment that lambing barns, corrals, pens, and feeding equipment it requires. Lambing barns also need more labor since you will need to make sure they are clean and dry. If lambing occurs during the winter, you must spend more on feed compared to spring lambing.

You will need to check on ewes and lambs at least every few hours and feed animals in each pen. You will also have to regularly change the bedding to maintain clean, dry, and comfortable surroundings in the lambing barn. A lambing barn should be free of the draft, so the risks of pneumonia are minimized.

### Pasture Lambing

You do not need a sizable initial investment or intensive labor for pasture lambing. Ewes can feed themselves by grazing, and pasture need not be cleaned, unlike lambing barns. However, pasture lambing is only recommended in temperatures above 45° F. Most newborn lambs can survive with sufficient natural shelter. Still, you will need to have plans to provide shelter in case of adverse weather.

Ewes are usually healthier when lambing takes place on pasture since they receive enough daily exercise. The chances of mismothering are also decreased since ewes have ample space to distance themselves from others while lambing. Pasture lambing also doesn't require you to check ewes and lambs after dark since ewes tend to lamb when there is some natural light available.

However, pasture lambing makes it difficult for you to offer assistance and treatment during lambing. Deworming, bottle-feeding, weaning, and record-keeping are also harder due to the difficulty of catching lambs on pasture. Newborn lambs also face significant threats from predators when they are born on pasture.

### Lambing Facilities

If you choose shed lambing, getting the facilities ready becomes as important as prepping ewes for lambing. The barn area should be clean with fresh bedding. You will also need to check for drafts and eliminate them beforehand. The drop area should at least have 12 to 14 square feet for each ewe.

It's advised that you set up lambing pens before the first ewe starts lambing. It's recommended that you have enough lambing pens for 10% of your flock. However, more is better since concentrated lambing is always possible. Smaller ewes can do with 4 ft. x 4 ft. pens, but larger ewes need 4 ft. x 6 ft. or 5 ft. x 5 ft. pens.

## Lambing Supplies

You will be more involved in the lambing process if you adopt shed lambing. In that event, you will require a list of supplies that will enable you to provide a helping hand to your ewes during lambing. Keep a stock of rubber or latex gloves for helping with difficult births and when handling newborns. Difficult births may also require OB lubrication, snare or leg puller, nylon rope, and disinfectant.

You may have to use a bearing retainer, prolapse harness, or ewe spoon to hold the vaginal prolapse in. Keep a warming box or heat lamp to warm up chilled lambs. It's also advised that you have the required antibiotics and needles and syringes for giving shots during and after lambing.

A thermometer comes in handy when diagnosing problems. Betadine or gentle iodine must dip naval chords. Some newborn lambs will require help feeding. Therefore, keep an esophageal feeding tube, frozen colostrum, and colostrum replacement handy. You're also going to need lamb milk replace, lamb nipples, and a lamb bar to feed multiple orphan lambs. You can also administer 50% dextrose for weak lambs.

An oral dosing syringe and an S-curved needle for suturing can also be handy. You can tag and maintain records of the newborn lambs during shed lambing. However, keep your choice of ear tags, an applicator, docking, and castrating tools, hanging scale, weigh sling for newborn lambs, and a pocket notebook for record-keeping. A head stanchion can also be helpful to graft lambs to encourage ewes to accept their own lambs.

# Dystocia or Assisting With Difficult Births

Difficult births are one of the leading causes of lamb death, according to experts. Various factors can cause dystocia in your flock, including miscarriage, malpresentation of the fetus, the disproportionate size of the lamb and ewe, cervix failing to dilate, deformed lamb, and vaginal prolapse. The biggest challenge you will face during lambing will be to determine when to assist the ewe or when to call for help.

You're advised to check on the ewe if she has been straining for an hour with no sign of the lamb. You must clean the ewe's backside and wash your hands using warm water and soap before entering her. Wear clean gloves whenever you examine a ewe and lubricate your hand all the way up to your elbow using a non-irritating lubricant.

Fold your fingers, creating a cone shape, and insert the hand into the ewe's vagina. You will feel the lamb's nose if the cervix is open. It should be gently resting on the front legs of the lamb. If the lamb is presented this way, the ewe should deliver without requiring your help unless the lamb is too big for the ewe's pelvic opening. If the lamb seems too big, gentle assistance is recommended.

Avoid pulling the hand in and out of the ewe and avoid changing hands without cleaning them again. You can also try to elevate the ewe's hindquarters or get her to stand up, so there is more room for repositioning. If you're unsuccessful in your attempts for half an hour, call the vet. It's highly advised against excessive pulling. Delayed delivery and excessive pulling can cause serious injuries to the ewe and lamb.

It's important that you never attempt to deliver a lamb when the birth canal isn't fully dilated since it can seriously injure the ewe. Lambs usually assist with their own birth to some extent. Be sure there are no more lambs remaining in the uterus after each delivery. It's advised that you provide a long-acting antibiotic injection to every ewe that you assist with delivery.

## Backward Presentation

Sometimes the lamb may present itself with its hind legs first. There is no need to turn the lamb since it can cause injury or death to the lamb and damage the uterus. The lamb can be born normally with some assistance. Backward presentation is common with twins and triplets.

## Elbow Lock

A lamb in the normal position can sometimes have its knees locked inside the birth canal. If you sense such a position, gently push the lamb back so the legs are extended.

## Leg Bent Back

If you feel that one of both legs of the lamb is bent back, gently reach the hoof and cup it in your palm. Then move it forward. If you can't straighten the legs, you may need to use a lambing rope on one or both the legs and push the head back so the legs can be straightened.

## Head Back

If the head is positioned back without resting on the front legs, gently push the lamb back. Then slowly turn the head to the correct position. It's advised that you attach a lambing rope to both legs so that you can locate them after pushing the lamb back. Don't pull by the lamb's jaw – which can injure it. Gently use the eye sockets for leverage to pull the head forward.

## Tight Birth

Disproportionate size causes lambing difficulties most of the time. You can assist by providing good lubrication and gentle yet firm pulling. Use the skin above the head of the lamb for leverage and extend one leg at a time.

## Breech

When a lamb is positioned backward with its tail near the opening and legs tucked under, it's known as a breech. Gently bring the lamb's rear legs forward. Provide assistance and ensure that the lamb is delivered quickly. The umbilical cord usually breaks before the lamb is born in this position so the lamb can suffocate if the delivery is delayed.

## Swollen Head

The lamb's head can become swollen if it has been outside the ewe's vulva for some time with the tongue sticking out. Lambs can usually survive in this position for a long time, although they may look cold and dead. Ensure that the head is clean by washing it with warm water and pushing it back into the uterus. Provide plenty of lubrication and determine the position before providing assistance in delivery.

## Simultaneous Births

Ewes belonging to flocks with higher lambing rates can often run into this problem; twins. These lambs will have their legs intertwined. You will need to first determine which leg belongs to which head. Then untangle the legs and push back one lamb so that the other has enough room to be delivered. Triplets are usually expected in simultaneous births.

## Dead and Deformed Lambs

Delivery and removal of dead or deformed lambs usually need the assistance of a vet. Such lambs rarely pass through the birth canal. Lambs dead for a while will need to be removed in pieces, and freshly dead lambs can be extracted normally.

### Ring-Womb

This condition takes place when the cervix fails to dilate. A caesarian section is usually required, as ring-womb doesn't respond to manipulation or medical treatment. You should not breed ewes that have experienced ring-womb before.

### Disinfecting Navels

Infectious agents can get into the newborn lamb through its navel. If the navel cord is over two inches long, you will need to clip it closer to the body. Disinfecting navel stumps soon after birth can avoid infections. You can achieve this by dipping or spraying the navel area with betadine or gentle iodine.

# After Lambing

You need not interfere with the ewe after a normal lambing since it can take care of the newborn lambs. Simply wipe away any mucus that may be stuck on the lamb's nostrils, and then the ewe will claim her lambs, allowing them to nurse. Most lambs will be up and nursing within an hour after birth.

### Colostrum

The "first milk" that ewes produce after lambing is known as *colostrum*. It contains high amounts of essential nutrients and antibodies key for the lambs' future health and performance. It's critical that the lambs drink enough colostrum within 18 to 24 hours from their birth, especially when it comes to receiving antibodies since they do not receive antibodies that are in the ewe's bloodstream before birth.

Thankfully, lambs can naturally absorb high amounts of antibodies during the first 18 to 24 hours from birth. You're advised to ensure that a lamb receives at least 10% of its bodyweight worth of colostrum during this time. Although lambs can survive without colostrum, the chances of disease and death are higher if they don't.

## Weaning Lambs

This is an important part of raising and breeding sheep. Weaning is the process of separating lambs from their dams, giving up their milk diet and adjusting to a plant-based one. It's usually a stressful time for both the lambs and ewes. Consequently, you need to provide an easy transition that causes minimal stress.

### Timing

According to surveys conducted by the USDA, the weaning age of lambs in the United States is usually around four months. There isn't an ideal time for weaning since it's determined by many factors, including the availability of pasture or feed supplies, facilities, and target markets. Some lambs are weaned earlier than four months while others are left to naturally wean, which usually takes six months.

### Early Weaning

This usually refers to weaning lambs anywhere between 21 and 90 days from birth. Lambs can be successfully weaned early, given they are consuming enough dry feed, preferably one pound every day, and drinking enough water. Size can also determine if lambs are ready to be weaned. Many sheep farmers wean their lambs either at 60 days from birth or when they reach 45 pounds.

Lambs weaned early can generally convert feed efficiently into lean tissue. It's more economical and beneficial to feed grain to lambs since the conversion of feed into gains is much greater than converting milk into gains. Early weaning also reduces the stresses involved with lactation of ewes, especially young ones.

Lambs weaned early can also be placed in dry pastures for finishing. It also allows farmers to sell culled ewes earlier, usually for higher prices. Early weaning also enables lambs to be marketed much earlier in the year when the prices are usually high. Contrarily, that does not mean weaning lambs too early is a great idea since it puts lambs and ewes under immense stress.

## Weaning Orphan Lambs

Aim to wean orphan lambs when they are 30 to 42 days of age or when they reach 25 to 30 pounds in body weight. Most farmers prefer abrupt weaning, although some farmers provide them with a diluted milk replacer. Dry lots are more favored when weaning orphaning lambs unless you have a high-quality pasture available.

## Late Weaning

You also have the option of letting lambs wean naturally. It usually takes them around four to six months from birth. Spring-born lambs usually take longer than winter or fall-born lambs to weaning naturally. You're also advised not to let spring-born lambs on pasture with dams until they are ready to be sold.

Late weaning is more natural and less stressful on ewes and lambs. The milk production of ewes naturally decreases by the time late weaning takes place, which decreases any risk of mastitis. Late weaning also allows you to fully use available forage by feeding it to lambs while easily managing both ewes and lambs as a single group.

Weaning naturally can result in ewes and lambs competing for forage, especially high-quality forage. The risks of parasitism and infection with worm larvae also increase with late weaning. Late lambing also requires you to castrate male lambs before they are three or four months of age.

## Weaning Environment

Lambs go through more stress during weaning than ewes. They are separated from their dams and required to feed themselves without relying on the milk diet they are used to. Accordingly, ewes should be taken away from the lambs instead of lambs being removed. Remaining in the same location can reduce the stress that weaning causes, as lambs already know where the water, feed, and minerals are located.

You're advised to keep ewes and lambs far apart so that they don't hear each other. Also closely monitor lambs during weaning because they are more susceptible to parasitic diseases and enterotoxaemia caused by overeating. Enterotoxaemia prevention can be accomplished by vaccinating against it when lambs are six to eight weeks old, and then following it up with a booster two to four weeks later.

# Chapter 10: Sheep Diseases and Healthcare

Many diseases can affect sheep and lambs. While we'll summarize some of the most common diseases in the United States, remember that certain regions or states may have different common diseases. Ergo, it's important that you seek advice from local sheep farmers and an animal veterinarian regarding diseases, treatment, and prevention methods to ensure that your flock remains healthy.

### Miscarriage

Miscarriage leads to the termination of pregnancy and losing lambs. It can also result in ewes giving birth to deformed or very weak lambs that die soon after birth. Although it's usual for miscarriage to occur in some ewes, you need to be concerned if your flock's miscarriage rate is higher than 5%.

Many factors can cause this issue; therefore, good management and hygiene are highly advised to protect your sheep. Providing antibiotics in late-gestation and vaccinating against Vibrio and Chlamydia before breeding can reduce the risks of miscarriage in ewes.

### Caseous lymphadenitis (CL)

This condition affects the lymphatic system in sheep that results in the formation of lymph node abscesses. CL is a highly contagious disease that severely affects the internal organs if untreated. There is a vaccine for CL that can decrease the number of abscesses, although it can't prevent the disease from infecting sheep. However, there is no need to vaccinate flocks without CL.

### Foot Scald and Foot Rot

These two diseases are the most common hoof diseases that affect sheep and have caused huge losses to the sheep industry worldwide. *Foot scald* takes place when the tissues between a sheep's toes are infected, while *foot rot* refers to the infection of the hoof's underlying tissue.

Both foot scald and foot rot can be very difficult to control and eradicate. The most effective measures include proper hoof maintenance or trimming, foot soaking, foot inspections, topical treatments, administration of antibiotics, and isolation of new or infected animals. Culling might be required if certain animals don't respond well to treatment for the protection of the flock.

Although vaccinations are available for foot rot, they do not cover all the strains of the disease – and they may not prevent the disease from occurring. Therefore, proper hygiene and management of your flock are advised regardless of vaccination status.

### Internal Parasites

The most common health issue that affects sheep throughout the world is internal parasites. Many parasites can infect sheep, varying by region, year, and farm. The most common parasites include flatworms such as flukes and tapeworms, roundworms or nematodes, and protozoa or single-cell organisms.

The barber's pole worm (also is known as Haemonchus contortus) causes anemia or blood loss and bottle jaw. Coccidia are protozoan parasites that damage the intestines and cause ill thrift (like "failure to thrive") and poor weight. It's important that you control internal parasites to maintain a healthy flock of sheep. It requires you to use a combination of treatment and management tools.

### Good Management

Parasitic problems can be minimized with proper management of your sheep and using common sense. Providing clean water, feed, feeders that are clean and constructed in a way to avoids easy contamination, avoiding overstocking pastures and shelters, and isolating newly-acquired sheep are some fine examples of proper management of sheep.

### Providing Clean Pastures

The pastures you provide your sheep should not contain worm larvae. Pastures that haven't been grazed for 6 to 12 months by sheep or goats, fields where silage or hay crop has been removed, land grazed by cattle or horses, and pastures rotated with field crops are examples of clean pastures ideal for sheep.

### Pasture Rest and Rotation

Many sheep farmers misunderstand how rotating pasture can help control or worsen parasitic problems. Rotating your flock using smaller paddocks increases the chances of them being exposed to parasite larvae. A pasture requires at least three months for the level of infectivity to become low. Therefore, you must provide enough rest for pastures, and pasture rotation will be fruitful only when such rest is provided.

### Grazing Strategies

Experts believe that around 80% of parasite larvae are found within the first two inches of grass. You are advised to avoid making sheep graze forages that are less than three inches in height. Experts believe that browsing (eating the leaves and young twigs of trees and shrubs) also leads to fewer parasitic issues.

### Alternative Forages

Experts advise that making sheep graze forage containing tannin-rich plants lower parasitic issues than sheep that graze grass pastures. Condensed tannins can deworm sheep and reduce the development of larvae in feces and the hatching rates of worm eggs. Forage species such as birdsfoot, chicory, trefoil, and sericea lespedeza are highly recommended as alternative forages.

### Nutritional Management

Sheep and lambs on nutrient-rich diets usually have better immune responses against internal parasites. Higher protein consumption has also been proven to help with parasitic problems in sheep, especially in ewes after lambing.

### Proper Anthelmintic Use

The use of anthelmintics (a group of antiparasitic drugs that expel parasitic worms and other internal parasites from the body by either stunning or killing them and without causing significant damage to the host – also called "vermicides") can help you control parasitic diseases. Though, they are mostly used to enhance treatment effectiveness and slow down the rate of drug resistance developed by worms. You also need to accurately measure the weight of sheep to provide them with the right dosage. Under-dosing should be avoided since it can backfire by making worms resistant to the anthelmintics used.

It's recommended that you use oral drenching when administering anthelmintics while delivering it over each sheep's tongue. It results in the close of the esophageal groove that bypasses the medication to the sheep's rumen. Anthelmintics are more effective when the sheep's gut slowly absorbs them, so you are advised to fast the sheep for at least 24 hours to enhance deworming efficiency - but the sheep should be allowed to drink water.

Any newly-acquired livestock needs to be dewormed with at least two-to-three different anthelmintics families. Moxidectin and Levamisole are preferred due to their superior potency. It's recommended that you release the dewormed sheep into a "wormy" pasture so that the medications can dilute any resistant worms that are present.

### Ovine Progressive Pneumonia (OPP)

OPP affects many systems in the sheep's body resulting in various symptoms. This viral infection is one of the most common causes of death in sheep. One common symptom of OPP is a hard bag. It's a form of mastitis where both sides of the udder are affected, resulting in decreased milk production (or *no* milk production) that causes the death of lambs. Ewes with hard bag are also usually culled since they can infect other ewes.

OPP is very difficult to eradicate – or even control – since there is no treatment or cure for the disease. Blood tests need to be carried out on ewes that show symptoms and need to be isolated or culled if they are found to carry OPP. Lambs need to be removed and isolated from infected ewes. In such cases, you will need to feed them with colostrum and milk that are heat-treated.

### Respiratory Disease

Pneumonia – caused by bacteria, viruses, and the environment – affects the digestive tract of sheep. Affected sheep often refuse to eat due to depression, while showing signs of respiratory distress that includes coughing. Respiratory disease can be treated using anti-inflammatory drugs and antibiotics, but there is no effective vaccine to prevent respiratory disease. You can minimize the risks of respiratory disease by providing sheep - especially those who are housed – with good ventilation.

## Scrapie

This fatal disease attacks the central nervous system of sheep, gradually developing over years before symptoms finally appear. Scrapie is usually transmitted during lambing via the placenta, and blood tests are usually recommended to identify sheep and lambs infected with it. The sheep industry has worked hard to eradicate this severe disease, and the USDA requires all sheep to have an ear tag of tattoo to assist in locating the live animals born where another animal was diagnosed with scrapie.

## Sore Mouth

This is the most common skin disease known to the sheep industry. Sore mouth is caused by a virus that belongs to the pox family with symptoms such as blisters and lesions on the lips, noses, mouths, and other body areas. Lambs and yearlings are more susceptible to sore mouth, while humans can also be infected with it. Sore mouth can be controlled using vaccines in your flock if there is an outbreak. Infected sheep can be treated with antibiotics and WD-40 sprays.

## Biosecurity

The measures taken to minimize or prevent exposure to diseases is known as *biosecurity*. It is highly recommended that you introduce strict biosecurity measures into the way you manage your flock so they remain free of diseases. Spreading of diseases can cause harm to sheep enterprises of all sizes, so, seriously consider using biosecurity methods - even if your flock is small.

## Acquisition of New Animals

Diseases can be introduced to your healthy flock accidentally when acquiring new livestock; and it's one of the most common ways that diseases spread from one farm to another. Sheep carrying diseases will often look healthy from the outside, so you must not underestimate the importance of taking strict biosecurity measures when purchasing new sheep.

First, make sure that you are buying healthy animals, and they come from a healthy flock. Always inquire about the farm's disease status and health program before purchasing animals. You're also advised to only purchase animals from reputable breeders, although they may be more expensive. Reputable breeders are more likely to have good health programs in place. Therefore, the likelihood of ending up with disease-carrying animals is very low.

Purchasing sheep from closed flocks – a group of animals that haven't been exposed to another flock for at least three years – is also highly advised. Experts do not recommend buying sheep from sale barns, stockyards, or public livestock auctions, as diseases can spread quickly among animals under such environments.

### Isolate Newly-Acquired Sheep

Experts recommend that you isolate or quarantine new sheep for at least a month to verify that they are free from diseases. Doing so greatly minimizes the likelihood of introducing diseases to your healthy flock through newly-acquired sheep. The quarantine area should be ideally 100 feet or as far away from your flock as possible.

It's advised that you trim the sheep's hooves while they are in isolation and inspect them for foot rot and foot scald. Soaking their feet in a 10% zinc sulfate solution is also highly advised. You should also deworm the animals with all three classes, so prevent introducing drug-resistant worms to your farm.

### The Risks of Sheep Showing

There are some risks associated with showing sheep at sheep exhibitions and shows. If you show your sheep at such events, avoid any direct contact with other sheep, sharing of equipment, feeders, and waterers. Remember to disinfect any equipment you borrow from event organizers or lend to other farmers during such events. Also, it's important to isolate your show animals for a month upon returning from sheep exhibitions and shows.

## Shearing

There's a chance of diseases spreading through shearing. Shearing equipment needs to be disinfected when animals being sheared belong to different flocks. Small-scale sheep farmers may have to take their sheep to shearing sheds, especially if they only have a handful animals. In such situations, quarantine your animals upon returning and inspect for any signs of diseases before releasing them to graze your land.

## Proper Management

You're advised to properly manage your farm no matter how small your flock may be. Poor management of your farm can lead to the spreading of diseases through rodents and other wildlife. Proper management of waste and secure storing of feed can usually help you avoid attracting rodents. Take steps to control rodents if there is a rodent problem on your property.

If you have cats in your farmstead, make sure that they are kept away from your grain and hay storages. Vaccinating and neutering cats is also highly advised, so the cat population in your property remains healthy and stable. Immediately remove and properly dispose of any dead carcasses to avoid the spread of diseases to healthy sheep. You should never feed dead carcasses to dogs or other animals or leave them out to be eaten by wildlife and rodents.

# Chapter 11: Selling Sheep Products: Wool, Meat, and Dairy

The purpose of raising sheep may be to make significant profits. Even if making profits isn't a high priority for you, having revenue from raising sheep will help you with the upkeep of your flock. Sheep farming ventures have the potential to make money by selling lambs, meat, wool, and dairy. You will need to market your sheep products accordingly; each product has its pros and cons.

### Lamb and Mutton

Meat that comes from sheep less than a year old is known as a lamb, while mutton comes from older sheep. Consumers prefer lamb since mutton tends to have a stronger flavor. Teeth are good indicators of their age if paperwork, tags, or tattoos are unavailable for confirmation.

The lower jaw of lambs usually has eight milk teeth. A yearling has two cut permanent incisor teeth while a sheep has two pairs. The front shanks of a lamb carcass usually have two break joints that are red, porous, and moist. Mutton carcasses have two spool joints, while a yearling's carcass features at least a single spool joint.

## The Demand for Lamb

According to the Journal of Food Distribution Research, the current per capita consumption of lamb is approximately a pound per person, per year. It's still considered a specialty product in the United States, with approximately 30% of citizens having never tried lamb. Only 24% of the population consumes lamb at least once every year – *much less* than countries such as Australia and New Zealand.

## Lamb Grading

Various standards are maintained when grading lambs. Lamb is graded according to the carcass, live lambs, feeder lambs, age, sex, and weight, and shrink or drift. These quality grades usually indicate the eating characteristics and palatability of meat. Prime, Choice, Good, and Utility are the USDA lamb grades.

## Carcass and Live

The yield of quality of meat is sometimes used to grade lamb. Fatter lambs are more likely to be graded as Prime. There's also a comparatively good demand for lamb graded as Good in ethnic markets, as they prefer leaner or lighter lamb. Lamb is given a yield grade standard according to the percentage of external fat found in the carcass. The leanest lamb is graded as 1 while the fattest is graded as 5. USDA grades live lambs with the same grades used for the carcass. Some states have different split grades for live lambs.

## Feeder Lambs

These lambs only weigh around 60 to 90 pounds and are mostly sold to grazers or feed lots. Feeder lambs are sold as large, medium, and small, determined by their weight and frame. Some states have unique grading standards for feeder lambs. These grades are becoming less important, according to experts since they are considered as potential slaughter lambs. Ethnic slaughterhouses now purchase feeder lambs to cater to markets that prefer lighter and leaner lamb.

### Age, Weight, and Sex

The age and weight of lambs when harvested varies. The average weight of lamb harvested in the United States is around 135 pounds, although weights vary between 30 pounds and 200 pounds. Lambs are harvested in the U. S. when they are between two to fifteen months of age.

Consumers rarely have a considerable preference for meat that comes from ewes, wethers, or rams. Meat from rams has a strong flavor that most consumers dislike, but some ethnic markets have a preference for meat from intact males.

### Shrink or Drift

The weight that lambs lose while being transported to the market is known as *shrink* or *drift.* Shrink is usually due to losing stomach contents during the first 20 hours after departing the farm. The body then compensates for the decreased water and feed by utilizing nutrients and moisture in the tissues that cause weight to shrink further.

According to experts, length and mode of transport, temperature, diet, and age of lambs are some factors that determine shrink. Lambs are also highly likely to shrink more if they are taken off feed the night before the day of sale. So, under those circumstances, you are advised to provide the lambs the same diet the night before and sort them on the day of sale.

# Marketing Options for Meat

There are many ways you can market your lambs. Lamb marketingmethods can be divided into two main groups as commodity and direct. It's also important to understand that sheep marketing methods and practices vary according to the region and the size of the operation. Because of this, you're advised to select the most suitable option when selling your lambs.

### Commodity Marketing

Most lambs are sold into various commodity markets that include public livestock auctions, buying stations, and abattoirs. Commodity markets are for producers who sell generic products in bulk and favor larger and low-cost producers, especially those whose farms are located closer to terminal markets. Lambs are slaughtered soon after purchase at terminal markets.

### Public Livestock Auctions

Most lamb producers sell their lambs at public livestock auctions at sale barns. Some sales grade lambs into large lots where lambs from different producers can be mixed in different lots. Some sales, on the other hand, sell each producer's lambs as separate lots.

Many sheep farmers prefer selling lambs at sale barns since it's a very convenient option. There are also more opportunities since most sales barns have sales every week. Payment is also usually prompt.

But there are disadvantages to selling lambs at sale barns. You won't know the price, and it can vary from one week to another, depending on the local supply and demand. Sale barns usually involve fees, commission, insurance, and yardage costs.

## Dealers, Brokers, and Order Buyers

You can also sell lambs to livestock dealers, brokers, or order buyers who act as middlemen. This method usually helps you save costs associated with selling your lambs at sale barns. The price is negotiated before the sale, and transportation is sometimes arranged.

### Abattoir (Slaughterhouse)

You also have the option to market your lambs directly to a slaughterhouse or meat processor. Lambs can be sold either on a carcass basis or live. Pricing methods also vary from forwarding or formula prices to spot cash prices. You can also benefit from value-based marketing and grid-pricing when you sell directly to a processor.

Value-based pricing involves pricing lambs based on the individual value of each lamb, while grid-pricing is based on carcass weight, quality grade, and yield. Lambs closer to the grid are valued higher, while those who "miss the grid" are valued less.

### Direct Marketing

When you sell sheep products directly to the consumer, it's known as direct marketing. Direct marketing of sheep products takes various forms, such as selling products at farmer's markets, selling over the internet, freezer lambs, on-farm sales, and selling products to retail outlets and restaurants. The main difference between commodity and direct marketing is the volume. The volumes involved in direct marketing are usually much lower than in commodity marketing.

Direct marketing of sheep products is advisable for small-scale farmers since profits are usually higher. At the same time, costs such as transportation, processing, and labor can also be high. Direct marketing may be the best option for you if you live near a city where people favor small businesses instead of products that come from commercial farms.

### Selling Carcasses

The most common way of direct marketing lamb is by selling whole carcasses or half-lambs. These are also known as "freezer lambs" since consumers usually buy a whole carcass or half of it and keep it in their freezer. You can sell animals by their hanging weight if you use a federally inspected plant to process the meat according to the customer's preferences – and at their expense. The meat will be stamped as "not-for-resale."

There is a higher demand for grain-fed lamb due to the milder flavor, while the market for grass-fed lamb is also on the rise since they are believed to be much healthier. Customer preferences also vary according to the age, size, and diet of the lambs. If you have a good meat processing plant nearby, selling "freezer lambs" is highly recommended.

### Farmer's Markets

A growing number of small-scale farmers are selling lamb at farmer's markets. Farmer's markets are becoming increasingly popular as consumers look to purchase local produce, preferably from smaller enterprises. If you wish to sell lamb at farmer's markets, you will need to process carcasses at a plant inspected by the USDA. The meat will come labeled. Some farmer's markets and states may require licenses and insurance to sell lamb at farmer's markets.

# Marketing Options for Wool

Selling wool is the primary income source of certain sheep-raising enterprises. If you raise wool sheep, selling wool can help you earn revenue to keep your flock going or even make profits depending on the size of your flock and the wool's quality. Some wool breeds can be sheared twice a year. The wool produced by certain breeds is also of high value. Wool harvested also needs to be stored, prepared, and packaged correctly to maximize earnings.

# Characteristics That Determine the Value of Wool

The wool value is determined by various characteristics such as the yield, crimp, fiber diameter, purity, color, staple strength, and length. The fiber diameter of wool, which is also known as its *fineness*, refers to the thickness or diameter of wool fibers, which, in turn, determines the thickness of the yarns it makes.

Different parts of a fleece usually have different diameters. Higher variation of wool fibers is undesirable with wool, with more uniform fibers being more valuable. As a result, the average fiber diameter is one of the key factors that determine the value of wool.

Wool fibers usually have a natural bend or waviness, which is known as *crimp*. Hair fibers usually have no crimp, and coarser wools have less of it. Fine wools usually have more crimp. Yield refers to the amount of wool that is left after washing, which is known as *scouring*. Lanolin or wool grease, dirt, dust, sand, and vegetable matter all amount to the "shrinkage" in wool. The yield of different wools usually varies between 40% and 70%, with bulkier fleeces usually indicating higher yields.

Colored fibers are highly *undesirable* in the wool market, primarily since they can't accept the dye. Certain niche markets, however, may have a high demand for colored wool. Some weavers and hand spinners who prefer to work with naturally colored wool may pay high prices for colored wool. Still, finding such consumers is difficult compared to selling wool in the commodity market.

The *staple length* refers to the length from the base to the top of an unstretched wool fiber. Wool with longer staple lengths is usually more valuable. Staple strength determines how well the wool can withstand cleaning and manufacturing; thus. stronger wools are valued higher.

### Marketing Options for Wool

Australia, China, and New Zealand dominate the wool market, with China also acting as the world's largest wool consumer. The United States – compared to these giants in the wool trade – is only a minor player. Super-fine-wool has the most demand globally as it is used to make high-quality clothing and fashion products. Ways you can market wool can be divided into two categories:  as a commodity and via direct marketing.

### Commodity

These markets are generally more suitable for large wool producers. However, that does not mean you can't market wool in the commodity market. You're advised to weigh your options and select commodity markets if you find them convenient and profitable.

## Wool Pools

Large producers of wool have enough produce to directly market to wool mills and warehouses. However, most wool producers, especially small-scale farmers, rarely have enough wool to do so. Wool pools are large stocks of wool that consist of contributions of multiple producers.

Countries, regions, and states operate them so there is enough wool to directly market to warehouses and wool mills. Wool is classed at wool pools before being sold according to their quality and type – and there has been a reduction in the number of wool pools in recent times. You can explore this option if there is an operational wool pool in your region.

## Wool Warehouse

Both private and cooperative wool warehouses play the role of the broker when marketing wool. The Roswell Wool warehouse in New Mexico is the largest wool warehouse in the U.S., while the oldest wool warehouse in the U. S. is Ohio's Mid-States Wool Growers Cooperative Association.

## Fiber Co-Ops

Cooperatives of varying sizes have been formed to maximize the value offered to wool by producers of different scales. These alliances usually have international partners and can market different amounts of wool.

## Direct Marketing

You can market wool directly to consumers in different ways. The most common direct marketing method is to sell whole fleeces to weavers, hand spinners, and other craftsmen. Different consumers usually have different preferences. Therefore, you need to understand what such consumers prefer before marketing your wool.

The preferences of hand spinners usually vary with type and color. Longwools are generally the most popular among them since they are easier to spin. Some hand spinners prefer more variety as they use different types and colors of wool.

Almost every hand spinner prefers clean wool, so, you'll need to *skirt* the fleeces (remove "junk" wool, second cuts, stains, or vegetable matter before processing.) It's common for some farmers to cover their sheep to make sure that their fleeces remain as clean as possible, which usually fetches higher prices.

You also have the option to market wool as *rovings*, which are made by washing and combing wool into twisted clumps. Rovings are used for felting, spinning, padding, stuffing, and crafts. You can also go a step further and make yarn from your wool. Some small-scale sheep farmers market many finished products made using the wool their sheep produce, including rugs, outerwear, garments, bedding, and more.

## Marketing Options for Seed stock

Every sheep-raising enterprise need not expect revenue from marketing wool, meat, and milk. You may raise sheep to breed sheep as a hobby or as a profit-making enterprise. There is a high demand for both purebred and crossbred seed stock. You can also market both registered and unregistered rams and ewes.

It's important that you do enough market research to identify breeds and breed crosses that are in high demand in your region. Certain breeds and breed crosses have established markets, while some may have niche markets. These markets usually vary geographically. Appropriately, you're highly recommended to do a proper market analysis before breeding, especially if you wish to earn an income by selling seed stock.

### Important Traits for Seed Stock

With dam breeds, fertility, early puberty, mothering ability, prolificacy, pounds of lamb weaned, milk production, easy-care, efficiency, resistance to foot rot, parasitic problems, and other diseases, fleece traits and longevity are key traits. Libido, lamb survival, lamb vigor, feed efficiency, carcass traits, post-weaning growth, and resistance against Scrapie are traits that are important for sire breeds.

### Breed Registries

Breed associations set the standards for specific breeds. Those standards determine the eligibility for registration of purebred sheep. Thus, breed registries are very important to production and marketing of high-quality seed stock.

### Standards of Health

Sheep farmers who sell seed stock must maintain high health standards so their animals are more desirable to other farmers. Your stocks need to be closed or mostly stocked for biosecurity reasons. Breeders are highly advised to enroll in the Voluntary Scrapie Flock Certification Program. It will help later by making your seed stock more attractive to potential buyers due to resistance against Scrapie.

Your seed stock should also be free from many other common diseases, especially foot rot, sore mouth, pinkeye, epididymitis, and caseous lymphadenitis. You're also advised to test your flock for caseous lymphadenitis, ovine progressive pneumonia, Q fever, and Johne's disease.

# Performance Record-Keeping

You are highly advised to maintain detailed flock records of your sheep when producing seed stock. Birth records, weaning weights, and the rate of gains after weaning are very important. You're also recommended to adjust weaning weights according to the sex of lambs, type of birth, rearing, and the dam's age.

If you produce terminal sire breeds, you're advised to collect carcass trait data. You can use ultrasound scans to determine features such as back fat and loin depth. Like other data, you need to adjust carcass data so proper comparisons can be made. Also consider having your flocked enrolled in the NSIP that calculates EBV, a measure of animals' genetic value.

You can also use Central Performance Tests as an alternative to NSIP. These tests are conducted by evaluating the performance of rams by bringing them to a central location. Central performance tests are especially recommended for terminal sire breeds such as Suffolk, Hampshire, and Texel.

### Advertising

An easy way of advertising seed stock is by putting up signs at your property's entrance and on your vehicles. Newspapers and magazines provide effective advertising options as well. Also consider web directories, a website of your own, and maintaining a presence on social media to market seed stock. Another common method to promote seed stock is by exhibiting your best animals at sheep fairs, shows, and festivals.

### Export

According to the United States Livestock Genetics Export, Inc. (USLGE), the export market for sheep genetics from the United States is growing. Yet, exporting seed stock involves a lot of work, especially in terms of animal health. Different countries have regulations and requirements you will need to abide by and fulfill. Diseases are the biggest barriers to the international trade of seed stock. Therefore, exporting semen and embryos are becoming increasingly popular.

# Marketing Options for Sheep Dairy

Although the demand for lamb and wool is much smaller than the market for sheep milk and dairy products, it is steadily increasing. People looking to eat healthily and purchase locally may buy sheep milk and dairy products from you. Sheep milk can also be marketed to local enterprises that produce sheep dairy products such as cheese, yogurt, and ice cream. You can use resources on the Specialist Cheesemakers Association website when you do market research before raising sheep for milk.

You might discover direct marketing options to sell cheese, yogurt, and ice cream to consumers themselves or restaurants that usually value quality over price. You might also find opportunities to market sheep milk and dairy products to local retailers if there is a market for them in your region.

### Marketing Methods

The market for sheep milk and dairy products is a niche market. It's recommended to maintain a strong presence online with a good website and social media pages. It's also advisable that you advertise on local farming and culinary media such as newspapers and magazines to give enough exposure to your products.

# Chapter 12: Showing Sheep

Showing sheep is great for those who raise sheep as a hobby and breed sheep. If you are passionate about sheep and showing sheep, take part in sheep shows, exhibitions, and festivals despite your experience or scale of the enterprise. Most principals of showing sheep are universal, although there can be certain regional principals.

### Maintain a Show Schedule

Preparation is a key ingredient of successful sheep showing. Know when the next event is and have a clear plan on how you will prepare for it. Maintaining a show schedule will certify that you prepare well for sheep shows. It makes tasks involved with sheep shows, such as washing, trimming, and transportation timing more organized and effective.

### Preparation

It's advised that you do a rough shear approximately two to three weeks before a show so that your sheep will be easier to wash and slick shear. Doing so will also make your sheep appear much fresher, cleaner, and attractive. Showing meat breeds require "cutting out" about a month ahead of a show. It involves slick shearing most of the sheep's sides and torso, so there is enough time for wool regrowth. It also makes final preparations much easier and quicker.

Card, curry, and finish trimming needs to be done a few times on the days before the show, so the wool has a firmer and smoother finish. Also arrive on time for sheep shows, and the hour before showing should be spent brushing, applying hide conditioner, and slick brushing.

Showing meat breeds require shearing the belly area around two weeks before a show. It ensures that your sheep look cleaner and more attractive in the show ring while allowing the wool to regain its color to provide a natural appearance. Blocking (a very close shear without the blades touching the sheep's skin) is also recommended a few days before a show.

Sheep breeds aren't washed for shows to allow the evaluation of natural wool. Still, cleanliness is very important when showing sheep, so pay attention to your sheep's ears, eyes, nose, and other areas, wiping them down to remove any debris or dirt before your class is called.

### Train Your Sheep

It takes time to train sheep for show. Most sheep farmers, especially beginners, use sheep halters to break and show their lambs. It needs to be done before the date of the show. Those who plan on showing sheep without a halter are still advised to train their sheep using a natural gait so they perform well in the show ring.

Also spend time training your sheep to be in a "Set Up" position. Ensure that your lambs are comfortable having their legs touched or re-positioned. It avoids incidents where lambs panic while in the show ring.

### Your Dress Code

Showmanship matters when showing sheep. It's recommended that you carefully check each show for specific dress codes, and wear appropriate attire. If no dress code is mentioned, dress professionally anyway; tie your hair up so that you can maintain eye contact with the judges. Pay careful attention to smaller details regarding your appearance.

### The Importance of Arriving on Time

You should not only avoid being late to your show class but also avoid arriving too early. Arriving late will make things stressful for you and interfere with your final preparations. Arriving too early will cause you to wait around with your lamb and create congestion at the ring entrance, resulting in your animal becoming tired and stressed. Hence, you are advised to check the show order for your show class in advance so you can arrive with enough time to be ready to enter the show ring five minutes before your class is called.

### Pay Attention to the Judge

Maintaining eye contact with the judge is something that many beginners forget to do. It allows you to be wary of the judge's position in the ring and the directions he or she may provide the showmen. It's sensible to avoid becoming too preoccupied with positioning your animal.

It's important for you to be aware of what is always happening in the show ring, including the judge's position in the ring and the positions of other animals and their showmen. Being aware of their positions will help you prepare to brace your lamb, creating a great impression of your animal and to avoid missing any directions provided by the judge.

Paying close attention to the judge's position is paramount in showmanship classes to ensure that your movements allow an unobstructed view of your animal. Finally, maintaining eye contact and putting on a smile can go a long way towards helping you – and your animal - make a good impression.

### Setting Up and Bracing Your Lamb

You will be asked to line up so the judge can evaluate the animals in the ring upon entering it. Different judges work differently. Watching a few classes in the event before your class is called is important, giving you a clear idea of how animals are set up and evaluated. Most judges are consistent with the way they do things, so being familiar with how they work can help you set up and move your animal most pleasingly.

### Setting Up Lambs for Shows

When showing sheep, positioning their hooves so they look more balanced, natural, and squarer while standing is known as *"setting up."* It's important for you to realize there isn't a "correct" position for all sheep. Some lambs may look more attractive with their feet set up wider in the rear, while some may look better with their feet stretched out. You can get advice from 4H leaders and experienced showmen at shows and fairs. You can then practice those tips on your trimming stand as you prepare for the show.

Generally, most sheep are set up with their feet sitting squarely under the corners of their body. The hocks on the rear legs need to be half an inch beyond a perfectly vertical position. The front legs should be positioned squarely under the shoulders. The sheep should appear straight when the judge evaluates both from the front and rear.

The animal's head should be set up in a natural-looking position with the nose pointing forward. Place your hand under the sheep's chin extending towards the neck. It allows you to hold the head high while checking that the nose comes down just enough so the jaw remains level with the back.

### Bracing Your Lamb in the Show Ring

The judge will handle your sheep at least once to check the animal's conditioning, thickness, length, and structure. Proper bracing makes certain that your sheep can be comfortably handled, giving you an advantage over your competitors. The best bracing position requires you to position your knee against the animal's chest while gently pressuring the animal's head held high. Doing so encourages the animal to slightly push back against you while flexing its muscles.

Practice bracing before attending sheep shows, and get help from another person. Your lamb must be comfortable with bracing so it doesn't become frightened when the judge eventually handles it. A frightened animal is very difficult to judge, and the results rarely go in your favor.

Beginners take time to learn how to brace animals. With experience, you can get any sheep to brace correctly. NOTE: You're highly advised to avoid lifting the lamb or twisting its neck in attempts at bracing.

Here's another book by Dion Rosser
that you might like

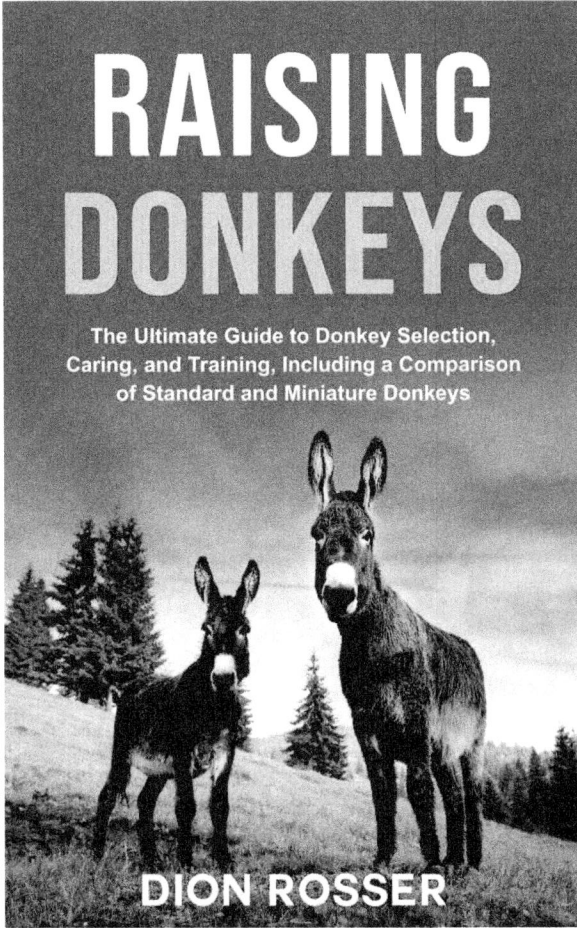

RAISING
DONKEYS

The Ultimate Guide to Donkey Selection,
Caring, and Training, Including a Comparison
of Standard and Miniature Donkeys

DION ROSSER

# References

*5 Questions to Ask Before Keeping Sheep.* (2016, March 1). Hobby Farms. https://www.hobbyfarms.com/5-questions-to-ask-before-keeping-sheep-3/

*6 Reasons To Raise Sheep • Insteading.* (n.d.). Insteading. https://insteading.com/blog/why-raise-sheep/

*13 tips for lambing outdoors.* (2020, January 4). Farmers Weekly. https://www.fwi.co.uk/livestock/husbandry/livestock-lambing/13-tips-for-lambing-outdoors

*A beginners guide to lambing...* (2015, February 26). Indie Farmer. https://www.indiefarmer.com/2015/02/26/beginners-guide-to-lambing/

*Breeding Sheep - What You Need to Know.* (2018, July 23). Timber Creek Farm. https://timbercreekfarmer.com/breeding-sheep-on-a-small-homestead/

*How to Raise Sheep on a Small Acreage for Profit.* (n.d.). Small Business - Chron.com. Retrieved from https://smallbusiness.chron.com/raise-sheep-small-acreage-profit-55996.html

*How to Show Sheep: 7 Sheep Showing Tips for Any Level.* (n.d.). Raising Sheep. Retrieved from http://www.raisingsheep.net/how-to-show-sheep.html

Mckenzie-Jakes, A. (n.d.). *02 Fun Facts About Sheep Fact Sheet II.*

*Pasture Vs. Shed Lambing.* (n.d.). EcoFarming Daily. Retrieved from https://www.ecofarmingdaily.com/raise-healthy-livestock/other-livestock/pasture-vs-shed-lambing/

Serban, C. (2018, November 22). *10 Most Popular Sheep Breeds raised for Meat, Fiber and Dairy.* Seradria. https://seradria.com/blog/most-popular-sheep-breeds.html

*Sheep 101 Home Page.* (2019). Sheep101.Info. http://www.sheep101.info/

*Sheep Shearing: How to Shear Sheep (Beginner's Guide).* (2018, May 8). ROYS FARM. https://www.roysfarm.com/sheep-shearing-information-guide/

*The Different Breeds Of Sheep - TheSheepSite.com.* (n.d.). The Sheep Site. Retrieved from https://www.thesheepsite.com/focus/5m/87/the-different-breeds-of-sheep-thesheepsitecom

*Welcome.* (n.d.). 2020 Scottish Smallholder Festival. Retrieved from https://ssgf.uk/exhibitors/beginners-guide-to-showing-sheep/

Printed in Great Britain
by Amazon